MW01625678

A PASSION OBSERVED

A PASSION OBSERVED

A TRUE STORY OF A MOTORCYCLE RACER

GEORGE JONAS

Macmillan of Canada
A Division of Canada Publishing Corporation
Toronto, Ontario, Canada

Canadian Cataloguing in Publication Data

Jonas, George, date.
A passion observed

Includes index.
ISBN 0-7715-9206-X

1. Mrazek, Frank. 2. Motorcycle racing - Canada.
3. Motorcycle racing - Czechoslovakia.
4. Motorcycle racing - Biography. I. Title.

GV1060.2.M73J66 1989 796.7'5'0924 C89-094107-6

Design: Don Fernley

Macmillan of Canada
A Division of Canada Publishing Corporation
Toronto, Ontario, Canada

Printed in Canada

CONTENTS

Foreword to Racers

Many track enthusiasts are busy improving their lap-times and consider other things only a waste. It's fair to warn them that this book is mainly about people. Hard-core riders may find much of the racing described here old hat. (In fact, they should.) If they're also hard-core readers, though, they might enjoy the story.

The events relate to an interesting period of the sport, from 1950 to the present, but this book is by no means a history of the period. It doesn't even scratch the surface. I hope someday somebody will write a comprehensive book about post-war motorcycle racing because it's a fascinating subject.

To novices, I'd like to point out that the techniques described in cornering, passing, or anything else are simply my opinion or the opinion of other people. Racing is a complex and dangerous sport. Competitors must gain their own experience, and the best place to start is a properly licensed racing school.

Good luck.

G. J.

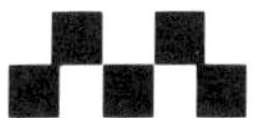

ACKNOWLEDGEMENTS

I'm grateful to Frank and Jana Mrazek, as well as to their daughters Sabi and Puppy, for having told me their story as they remembered it. Though I've relied on their recollections, the responsibility for any error or misinterpretation is mine.

I wish to thank the sponsors of Classic Racers Canada: Labatt's, and Messrs. Jim Allen, Monty Beber, Ted Burnett, Eddy Cogan, Edward L. Greenspan, Q.C., Peter Israel, Norman Jewison, Jack Matthews, Sam Meister, Sam "Stubby" Steinbaum, and Phil Turk for making possible the events described in the last chapter of this book.

Thanks are due to my patient editor, Guy Gavriel Kay, who is not normally called upon to deal with motorcycles in his otherwise exemplary life.

I'm much indebted to members, competitors, and officials of the American Motorcyclist Association, the American Historic Racing Motorcycle Association, the Ducati Owners' Club of Canada, Racing Associates Canada Events Inc., the United States Classic Racing Association, and the Vintage Road Racing Association of Canada for answering my questions, and especially for letting me hold up traffic at various racetracks in Canada and the United States while researching this book.

It is with our passions as it is with fire and water,
they are good servants, but bad masters.
— SIR ROGER L'ESTRANGE

A passion observed ceases to be a passion.
— BARUCH SPINOZA

THE PRIVATEER

I have spread my dreams under your feet;
Tread softly because you tread on my dreams.
Y. B. YEATS

The hillside is dotted with small fires. The mosquitoes—well, somebody's forgotten to show the mosquitoes the calendar. By August in these parts they're not supposed to be still buzzing around, so they're a month or two behind schedule. Big mosquitoes, nosing over for the kill. Coming in like dive-bombers. With real teeth, or stings or pincers or whatever the hell they use.

Actually, the mosquitoes are no problem. It's even possible to play a game with them in this fading light. The insect repellent works. Not exactly as advertised, because if you spray it on your skin the mosquitoes will just lap it up, but it's a great anti-aircraft weapon.

Here's what you do: you take off your shirt and sit on the ground, leaning against a rock. Can of spray at the ready, finger on the button. Imagine it's the Second World War and you're a Bofors battery, waiting for the Zeroes. The lead mosquito smells you and starts revving up the engine. Soon it's coming in, spinning about 20,000, until you think that it's going to throw a rod. You can't see it—or *her*, they're supposed to be females—but you can figure the direction from the sound, more or less. That's what makes it fair. So, when she's about a foot away, you let her have it. Zap! Right between the eyes.

It works every time. You can't see a thing, but if you've got the nozzle head on, it's like touching the kill-button. The sucker will cut the switch and drop like a stone. If you miss, she'll catch a whiff of the vapour and just motor away.

So the mosquitoes are no problem while the can of spray lasts and you stay awake. Sleeping could be a problem. The mosquitoes will get into the tent. Sometimes they even get into the sleeping-bag, but tonight it's too hot for a sleeping-bag anyway.

The small fires on the hillside are other privateers, barbecuing steaks or wieners. The veterans, the real connoisseurs, are probably barbecuing sausages. There's zilch catering at this track, but the sausages in the village are great. They're almost as good as the hot sausages at Daytona, and the veterans know it.

The people around the fires are all privateers, along with their wives or girlfriends. Factory guys with their wives or girlfriends would be sleeping in a motel. At Daytona they'd be sleeping in Daytona Beach or maybe in Ormond Beach. At Loudon they'd be sleeping in Concord, New Hampshire. At Road America they'd be sleeping in Elkhart Lake, Wisconsin. At the Shannonville track in Canada they'd be sleeping in Belleville, Ontario. Here at Grattan Speedway, if there were any factory guys here in the first place, they'd be sleeping in Grand Rapids, Michigan.

It's nice to sleep at the President's Motel in Grand Rapids. It's very comfortable. Double beds, swimming-pool, air-conditioning, no mosquitoes, breakfast included. Trouble is, you've got to be Kenny Roberts to afford it. King Kenny can probably pay 40 bucks a night. That's 80 bucks a weekend, or 120 if he wants to practise at the track a day before the heat races.

Maybe Kenny doesn't. He doesn't need the practice.

But you'd better be a world champion if you want to pay $120 just to sleep, because you still have to pay competitors' entry fee—say, $90 for a couple of classes—and maybe $50 for a five-gallon can of racing fuel. Then you'll have to fill up the van three times coming and going, that's six times 45; make it an even 300. Make it about $600 a race.

That's without a spark plug. Without tires. Without a quart of oil. Hell, that's even without a roll of duct tape.

Everybody has to fill up the van but everybody doesn't have to sleep in a bed. Everybody hasn't got $600 to spend on a race, which is why there's a good chance that a lot of bodies will be camping. You, for instance. Which is not a complaint, only a fact. You don't complain because—well, you don't complain for several reasons. First, at the track nobody complains. Who's there to complain *to*, when everybody's in the same boat? Privateers are, anyway, which means about ninety percent of everybody.

You don't resent the factory guys or the handful of privateers who have rich backers or sponsors. Hell, if liquid sponsors were falling from the sky you'd be out there to catch them in a bucket. For a backer, you'd drink battery acid and lick the spoon. You're not jealous of the guys who've landed a sponsor: you wish them good luck. You just keep dicing with them. That's what they're on the track for, aren't they? When the flag drops, the bullshit stops.

So, you reel them in whenever you can. Try to pull them for a lap. Let them breathe a little smoke. Let them see your front wheel now and again.

It's amazing, when you think about it, how often you've reeled in those guys with the big bucks. The big bucks, the big sponsors, the big trailers, the big motorhomes. The guys who have managed to catch the eye of some promoter, someone who makes all the difference. Someone who's never given *you* the time of the day. But those guys have somehow caught his eye, at least once, at least for a season or two.

You've smoked guys on the track who now own thirty-foot six-wheelers with captain's chairs and room for a kitchen and an air-conditioned repair shop in the back. Shelves for tools on both sides. Mounds of spares. Compressors and torches. Big teams who wrench under striped awnings on smooth outdoor carpeting, while you're groping for nuts and bolts by yourself in the grass with a wet towel on your head.

And you've reeled those guys in, you've motored by them. Not because they weren't pretty fast but because you happened to be faster.

Only maybe you didn't reel them in when it mattered. You'd pass them in the heats, then cook your clutch in the finals. Break the lap record, then DNF with a nickel's worth of copper coming loose. Lead in the regionals, then crash on the last lap. Bounce seven times, then slide on your ass for a mile.

You got the attention. You got the applause when you picked yourself up and waved to the spectators. All the other guy got was the chequered flag, the championship, the sponsors, and the factory ride.

Well, that's racing.

You don't complain, because nobody but nobody ever twisted your arm to race motorcycles. If anybody ever twisted your arm it was for you *not* to race, starting with your mother and your father. And as everybody—well, maybe not everybody, but as half the guys at the track will tell you—their wives, even if they were pit popsies to begin with, even if the idea that made them moist and curly around the nipples was to go out with a racer, even *those* girls by the time they were married for a couple of seasons started asking if it would never end. Started asking. Ended up demanding.

So, if it's tough going, there's nobody to blame. Nobody pushes anybody to race.

There are fewer mosquitoes around those fires on the hillside. For one thing, the smoke keeps them down, and there's also the breeze. Here in the hollow of the infield everything is still. The air is so heavy with moisture the bugs slicing through it sound like submarines. You can hear their wings splash. But it's the nearest patch of grass to the grid. The ground is not too soft, but it's soft enough. There are no stones. It's the ideal place for a guy with no crew to help him push the bike around, especially if he hobbles a bit. The last thing you want to do is stumble when you're bump-starting the damn thing. Plastic windscreens can crack as soon as you look at them, and they cost the earth.

The hobbling, well, there's nothing anybody can do about *that*. Not until after the end of the racing season anyway. It's a bit of bone, according to the doctor, growing around the plastic ball, the one that replaced the stainless steel ball, which was the one that replaced the joint.

It was a bad break, but even that's nothing to complain about when the other breaks had been pretty good. There's bound to be one bad break among all the good ones. If you race long enough you'll crash, and if you crash often enough you'll fracture something. If you fracture enough things, you'll fracture one thing like hell. The alternative is not to crash, but if you never crash you haven't been going fast enough and if you don't go fast enough it's a better alternative not to race. Which is impossible.

Right, Doctor?

The doctor has made a good living out of you these last twenty-five years and he'll be the first to admit it. "Give me more customers like you," he'll say after every visit, "and I'll be a rich man." Then you both laugh and you mumble something about trying to do your bit for the doc's kids at college, except you've got to be careful when you laugh because the pain's making you wince.

The doctor is happy to see you, and not because of his fees. It would be harder to say why he likes you. Perhaps it's because of the challenge. Not some real high-tech challenge because, thank God, you've never injured anything big like your brain or your spine, but he must be clever to patch up those broken collarbones or dislocated thumbs quickly enough for you to make the grid next week. Keep down the swelling and pain, make at least two of the five fingers mobile. No reason why the cast has to go directly on the foot when it can go over the racing boot. Pump in a few more cc's of cortisone, and away you go. That's the kind of thing sports doctors know about.

It's probably a relief for the doctor, too, after all those malingerers who show up at his office worrying about their blood pressures or head colds. It must be a relief, judging by his standard wisecrack as you're about to leave. "Whoa, what about a certificate for the boss?" the doctor will ask. "Or aren't you the type to take the day off because you've got a runny nose?"

You're old friends, the doctor and you. At least he considers himself a friend and he's probably right. Once the two of you even went out to dinner, together with your wives, and it was the doctor's idea. A pretty fancy restaurant downtown that took about twenty minutes to find even though he gave you directions. Dinner was steaks and a tumbler of old wine that had cobwebs around the bottle when they brought it up from the cellar. Then, about six months after that, you asked them to your home for dinner and your wife really put on the dog as she usually does. So now you're friends, not just doctor and patient, but when you look at each other it's still through the wrong end of a telescope. Maybe you envy him a little because you know that he's smart, and he envies you a little because he thinks that you're crazy.

What's there to envy about a crazy man? Probably that he isn't too scared. Most people are, and being too scared must be a heavy load.

And what's there to envy about someone who's smart? That's easy, it doesn't even need an explanation. That's why crazy people hang around with smart people and smart people hang around with crazy people sometimes, each hoping that some of it will rub off. Except it almost never does.

You don't know if the doctor got any crazier—perhaps he did—but you sure as hell never got any smarter. If you had, you wouldn't be sitting here zapping mosquitoes. You'd be lying at home on the sofa, watching the races on TV. Because the one thing you ought to know by the time you've turned thirty or forty—to say nothing of fifty—is that racing will never get you anywhere. For you, there won't be a big time. No easy street. No factory rides. No thirty-foot trailers.

A privateer past thirty knows that all racing will ever get him are more debts, more mosquitoes, more nights of listening to the whine of the rear end in a rickety van, and maybe a few more cracked ribs and separated shoulders. This much he knows for sure, which leaves him with nothing to worry about except whether his tie-downs from the season before will outlast his marriage or whether it'll be the other way around.

Yet, as it happens, it's privateers past thirty who fill up many spaces on the starting grid in the regionals, and in some of the nationals, too. Two-thirds of the spaces, in the vintage or the twin-cylinder classes. Not just privateers over thirty, but over forty and fifty. Even over sixty, one or two of them. No fool, as they say, like an old fool. Which is exactly what they say about you, right to your face, and you grin as if they were handing you a compliment.

You grin, sheepishly but proudly, the same way you grin at the doctor or at the nurses in the emergency room if you happen to be conscious when they bring you in. The way you grin when you display a bruise or a scar in the paddock. (Maybe not right after it's happened because then you still feel too furious or stupid, but a week or a month later.) You grin because you know, as do the guys in pit lane, that you've been to the edge. You've taken your skin to the edge and you've brought it back. You've touched the edge and you've brought back a mark. A smudgy fingerprint of the Grim Reaper. Which is as much as anybody can do.

But the funny thing is, you don't really feel crazy. You've always felt pretty sane, or even cautious and conservative (except a few times when the adrenalin started flowing on the track and you did something that later you thought was, well, maybe a little crazy). You simply accept that you're crazy because people have called you crazy all your life. Some with sneering disdain (which is fine; it's water off a duck's back), some with admiration, and some with a mixture of both. So okay, maybe that's what you are: crazy. If most people are sane, you have to be crazy because most people don't race motorcycles for a living.

If anything it's that phrase "for a living" that makes you wonder sometimes if you really are crazy. Because you can put two and two together as easily as the next person. You know that a little wooden plaque with a piece of engraved metal is worth twenty or thirty bucks, and a really elaborate cup is worth maybe a hundred or two, max. And you know that if you finish in the top three, what they'll give you is one of those little wooden plaques. You have two to three hundred of them in the basement by now.

As long as you're an amateur, a trophy's all they're going to give you. When you become a pro, you may also get a cheque. (In the rich classes like Superbikes, that is, because in a lot of classes even the pros get nothing.) In the featured classes the cheques that go with the trophies range from about $25 to $500 at most levels of racing, from club to national. A series championship is worth about $250 to $2,000. Add manufacturer's contingency (if any), add a share of the sponsor's or advertiser's purse (maybe). Add some start money (pretty rare these days). Add anything else you can think of, and a great season of finishing every race in the top three may net you $5,000, with luck, in a regional series. Triple that for a national series (but there you're not likely to finish in the top three so often).

The cost of that season, if you already own your racing bike, spares, tools, van, or trailer, will be about $15 to 25,000. Make it $25 to 50,000 if you need to buy more crankshafts or gearboxes, rent vans, or pay fabricators and tuners. In other words, a loss of $10 to 30,000 after a winning season.

A loss, if you have sponsors. A debt, if you're on your own.

That leaves the Internationals and the Grand Prix circuit. There's real money there, except privateers don't see much of it. Privateers rarely come in the top three in world class racing—not unless the factory teams blow up. In Grand Prix racing it's a red-letter day for a privateer to finish in the top fifteen. Which is great if the privateer has rich backers: then it's just a paid holiday. Why not ride around the Salzburgring in Austria, Hockenheim in West Germany, Le Mans or Paul Ricard in France, Brands Hatch in Britain, Jarama in Spain, Monza in Italy, Brno in Czechoslovakia, or maybe Anderstorp in Sweden. Get a thousand pounds' start money. Come tenth or nineteenth or twenty-third, meet some old pals, stay in a nice hotel, do a bit of bench-racing at the bar. Great stuff if you can get it. That's why it's useful to carry a bucket, just in case. One day sponsors may start raining from the sky.

There's no trace of rain tonight. The fires are flickering on the hillside. Some people are sure to be roasting wieners up there, maybe even spicy sausages. If you had some yourself, or if you could take along a case of beer, you could go sit by the fire. Fewer bugs, and one of those sausages would hit the spot. Hell, you could go sit by the fire anyway: at the track just about anybody would offer you a sausage or a beer. Maybe you wouldn't think twice about it if only you had the price of a case of beer in your wallet. But this weekend it's either another gallon of Cam 2 or a case of beer, and since the bike doesn't run on beer, the choice is pretty simple. So you stay where you are.

Zap another mosquito.

It's stupid, getting hung up on the question of money. It misses the point. Racing's a sport like any other. "Amateur" or "professional" are just words. They only mean a ranking, like brown belt or black belt in judo. In the old days they used to call it senior and junior, or sometimes expert, intermediate, and novice. Some places they still do, and maybe it's more accurate.

Most people *pay* money to play a sport. They pay a fancy club for letting them chase a little white ball across a green or swing at a bigger one on a court. Why should you have your fun for nothing?

But racing's a spectator sport. Isn't a spectator sport more like show business? Even at club level, hundreds of spectators are coming to the races. Promoters charge a lot of money to those people at the gate. Spectators pay to see you race, so shouldn't you get some of that money?

Would a TV actor perform for nothing? Well, yes, some probably would. Maybe they all would if nobody paid them. Just as you go on racing.

Just as some other guys do here at the track, the slow guys, who don't even get little wooden plaques. Guys who don't ever finish first, second, or third. Backmarkers, who think they've had a good weekend when only the race-leaders lapped them, and a spectacular one when nobody did. Mid-pack sluggers you rarely see after the flag drops, who spend the season dicing with each other. Yet those guys also have to fill up the van. They spend just as much for tires and spark plugs. They pay the same entry fees.

They also have to break a collarbone now and again—so count your blessings. At least you're not a slow guy. God alone knows what makes *them* go on racing. You've told yourself a thousand times that you'd quit the minute you stopped winning; that if you were "out of the wood", as they say, even at club level, you'd quit. You'd hang up your leathers if you were trailing even in the regionals. And then you worry that you may be lying because perhaps you'd hang up nothing, just as these backmarkers never quit.

Yeah, but they have good jobs, some of them. A lot of these slow guys are rich. They own their own businesses. They're in engineering or advertising or whatever. They can afford to drop a bundle. Their wives don't nag them when the bills start coming in at the first of the month.

Yeah, well, how could you tell what other people's wives nag them about? You don't have a tape recorder in their bedrooms.

Some of the fires are going out now. Laughter drifts over from the hillside, a couple of girls have started giggling. From Corner 8 it's the aroma of spicy sausages wafting across the infield, you just wish the mosquitoes would go for it. But they seem to prefer human blood. The spray-can's nearly empty.

Time to put up the tent.

Tomorrow's going to be a scorcher. The sun will suck up the heat from the tarmac below. The heat, and the grease, too: the old oil spills mixed with melted flecks of rubber. The track will be as slippery as hell. You'll have to look for someone to sprinkle talcum powder in your gloves before you put them on, you'll never get them off otherwise. Yeah, and maybe go down on the jetting after practice or else by noon the engine will be running too rich for the finals. Or better leave it, the jetting's perfect now, why risk melting the pistons? Run it cool and rich—but maybe a tad deeper into the corners. It's club-level racing. Power's not the only way to smoke those turkeys.

Some guys have crews. Crews to worry about their gloves or their jetting. They have groupies and mechanics and tuners: all they have to worry about is the race. They're the same guys who'll sleep at the President's Motel tonight. Well, good luck to them. Nothing counts anyway, except tomorrow. Those twelve minutes between the white flag and the chequered flag.

Sleep for seven hours like a baby. There are some riders who close their eyes then cut a dozen laps around the track in their minds. They say it makes a difference. They say if they concentrate before a race their lap-times go down by a full second. Good luck to them, too—though it beats you how planning can make a difference. What is there to plan for, your lines? Your lines depend on traffic. Are you going to queue up behind some guys because they happen to cruise in your perfect line? What's there to concentrate on when you've no idea what's going to happen on the track, when no two races are ever the same? Better to relax. Sleep for seven hours like a baby.

Zap! One of the buggers got right into the tent. To each his own, but these guys with all this natter about concentrating are ridiculous. Mike Hailwood said it all when they asked him how to go faster than the other guys—"Mike the Bike", perhaps the greatest racer ever, cut down by some highway drunk when he drove for a hamburger after a meet—Mike just told them, "More gas, less brake."

But maybe these people are nervous, that's why they're concentrating so much. Maybe they're scared. They'd like to win a race without dropping the bike, so they're forever drawing diagrams and watching videos about lines and apexes.

It's not too hard to win in a diagram. There's an easy way never to drop the bike: stay home.

Staying home is plain smart. Your wife is probably at home now, with the phone close to her bedside. For some years she hasn't come to the track. It's Saturday, so chances are she's watching a rerun of "Dynasty". She could watch one of those shows every night. A woman of many interests, reads magazines, listens to music. She can talk about anything, not just racing. Sharp dresser, too, right at home in the fanciest places; whatever made her pick a grease monkey like you. Never mind, she has her regrets by now.

A breeze has started up, it's fingering the ropes around the tent. Funny, the wind usually dies down at night, maybe the weather's going to shift. Some fires are still flickering on the hillside. Yes, she's probably watching television in the bedroom.

2

THE FLAG DROPS

Motorcycles didn't run in the family.

Frank Mrazek's father used to be big on soccer. As for his younger brother, Jiří—just a kid, born in 1942—he was playing on the junior hockey team of the Police Club. Hockey was the thing in Brno in the early 1950s, and gifted juniors in the Police Club could look forward to a professional career in the rink. Nominally they'd have to join the police force, of course. In the Socialist Republic of Czechoslovakia all sports were amateur. Or all sports were professional, depending on how one looked at it. A hockey star in the Police Club would be issued his uniform, his rank and his salary like everybody else, and the only thing he wouldn't be issued was a duty roster for walking a beat. So some policemen walked and some policemen skated, and what was wrong with that?

Old man Mrazek wasn't really part of this new world, though he tried his best to accommodate it. Czechs like František Mrazek, Senior, had long elevated the art of accommodation into something close to a science. Not by choice either, but through the main force of history. Other nations could look to role models like the Knights of the Round Table, but Czechs could only look to the Good Soldier Švejk. Except that accommodation wasn't easy for someone like Mrazek, Sr., a rotten capitalist. Or a "kulak", as they called it back then.

What a kulak was, was someone with a big yard and a big house in the back. A chestnut tree in the middle of the yard. Also a mill, a garage, a stable, and a shed or two attached for the trucks. The trucks were used for cartage and hauling, which was old man Mrazek's business.

Father Mrazek could do two things really well—three, if you counted soccer. First, he could watch his pennies. Second, he could fix engines. Any kind of engines. Cars, trucks, pumps, stationary engines for mines or factories, it was all the same to him. If they didn't work, he could fix them.

Motorcycles didn't run in the family, so what happened to young Frank must have been destiny. It was nobody's fault that the city of Brno (not one of the great cities of the world, really, latitude 49°18′ north, longitude 16°37′ east, about 180 kilometres southeast of Prague as the crow flies) happened to boast a popular racing circuit. Not man-made, only man-adjusted: a sort of East European Monaco or Spa Francorchamps, a true road circuit, running through cobblestoned streets and neighbouring villages at the outskirts of the town.

Nor was it anybody's fault that Frank Juhan, DKW works rider, European champion (and not a young man by 1947), came from Germany that year with his supercharged 250 cc DKW to this very circuit. He came, he saw, he conquered. Just like Caesar, though eleven-year-old Frank Mrazek knew nothing about that.

What Frank knew, as he was watching Juhan take the flag on the DKW, was: Okay, this is my life. This is it. This noise, this smell. Later he would remember very little about the race, why he went to it, how he got there, but he would always remember that moment. It was a revelation without any doubts or questions. Nothing mystical, only sensible. Like stepping into a dark room and flicking on the light.

It was nobody's fault. Not many people decide their entire lives in a single moment but some do, and Frank was one of them. He knew that he was put on earth to race motorcyles.

He knew that he could ride them faster and smoother than anybody else in Brno, maybe in Czechoslovakia, maybe in the world. He knew it before he'd ever swung a leg over one.

This—if you stop to think about it—is the only tough decision anybody ever has to make. The rest is just figuring out how, and that's easier. Most people would know *how*, if they only knew *what*. Frank knew it from the age of eleven.

Motorcycles didn't run in the family, but Frank's mother happened to have a brother and this brother happened to have a son. A first cousin, in other words, or some such thing—Frank was never much at working out exact degrees of kinship. In any event, first-cousin-or-some-such-thing Mrazek was racing a 250 cc motorcycle just like the German champion Juhan, though not an exotic DKW but a home-grown Jawa. So Frank started hanging around his cousin. By the time he was old enough to apply for a driver's licence (fourteen for small motorcycles) he had been to dozens of races. Helping to load bikes, pumping up tires, cleaning greasy parts in gasoline, fetching, wiping, carrying—and watching, watching.

There was still a problem. By the time Frank was old enough to apply for a driver's licence, he still wasn't old enough to receive one. Not on his own, anyway. He could receive a licence only if his father signed for it, assuming responsibility for all harm young Frank might cause to himself, to others, to the people's Republic of Czechoslovakia, to the socialist camp, maybe to all peace-loving nations. So there was a problem.

Two problems, really. The first problem was old man Mrazek's reluctance to sign anything that might cost him at the other end. Nobody who's good at watching his pennies likes doing that. But the second problem was more specific. It might have prevented even a generous person from putting his name on this particular document because the fact was, Frank appeared to be somewhat accident-prone.

For instance, there was that matter of a game of cowboys-and-Indians when Frank was about ten. By the end of the war all kids wanted to play cowboys-and-Indians, even in Czechoslovakia, and it was also natural that the kids liked to use the

Mrazeks' yard. It was the biggest one in the neighbourhood. But not all kids would try to jump from the roof of the house into the chestnut tree. Only Frank tried to do that, and he not only broke his leg but broke a big, healthy limb off the tree.

It was Mrs. Mrazek's favourite tree. Frank's parents weren't home when it happened, but he had to come up with the truth when they got back since they would have seen the damage anyway. They would have noticed a broken leg and a broken tree, and they were good at putting two and two together.

Father and Mother did end up taking Frank to the hospital, as all concerned parents might, the minute Mrs. Mrazek finished beating him with a horsewhip. But neither the whipping nor the doctors helped much, because soon after the cast came off the leg, another one had to go on his arm when Frank fell off a high railing. (Well, really, he was sitting on it and lost his balance; it could have happened to anyone.) Then, with that cast still on one arm, he tried to ride a small 100 cc Jawa trailbike in the yard, so a new cast had to go on his other arm as well. The last accident really wasn't Frank's fault, though, because it was Jiří who had stuck the handle of the spade through the spokes of the front wheel.

The Mrazek boys were a lively pair, and the only relief for František and Marie Mrazek was their normal daughter, Dana, who broke nothing (though even she ended up marrying a motocross champion). Not that Frank's father objected to liveliness. It was unruliness that he didn't like. His house-rules for adolescents were simple. No smoking, no drinking, no swearing, no damaging of chestnut trees. And definitely no talking back.

The punishment for breaking the rules was swift and direct. Perhaps Mother Mrazek was quicker with her hands, but Father Mrazek was heavier. The exercise of parental authority had no cut-off point. Frank was already a soldier on leave when his father smelled cigarette smoke in the house. Sniff. Did you smoke? No, I didn't. Don't lie to me. Wham!

Far from having anything against sports, the old man took them only too seriously. František Mrazek liked to play for keeps. He was not particularly big or strong, but people did not push him around on the soccer field. They knew that if they kicked him once, he'd kick them four times. Soccer is a fairly rough game, even amateur soccer, and though František didn't have the reputation of being a dirty player, he did have a reputation for being unforgiving. Whatever his standards were, he had them. They were his own. No one could violate them with impunity, on or off the field.

It was to such a man that young Frank had to turn for a signature for a licence. There's no telling how his future might have developed if Father Mrazek had refused, but he didn't. He grumbled, threatened, delayed, but in the end he signed.

Which left only SVAZARM to contend with.

Whatever the dread letters stood for (Frank could never say), by the 1950s it was SVAZARM—SVAZARM, the all-powerful, the almighty—that was in charge of sports in Czechoslovakia that had a possible military use or connection, such as shooting, flying, parachuting, scuba-diving, or racing motorcycles. In a sense SVAZARM was simply a paramilitary sports club, but since the state permitted no private clubs it became the sole body for sanctioning events and licensing competitors in a variety of sports. If you didn't play with SVAZARM, you just didn't play.

Back in those days a youngster in America would slap on some number plates, wire the drain plugs, then haul his bike to a racetrack that some local club or promoter rented for a weekend meet. He'd pay his entry fee, grin at the scrutineers (if any), sign a waiver (maybe), then push his bike on to the grid and race.

But this, even before and apart from communism, just wasn't the European way. To operate a racing machine on the track, you needed a racing licence. (You needed a licence to drive just about anything in Europe, including a push-button elevator.) And SVAZARM wouldn't give you a licence to race on the pavement until you first proved yourself in the dirt.

Maybe in the long run this wasn't such a bad thing. Trails may be tough to ride, but dirt is relatively soft. The pavement is unforgivingly hard. Riding enduros for about six months Frank learned to keep a little CZ rubber-side down, shiny-side up, most of the time. He also collected enough points for his licence as a novice road racer. Now he had the documents that entitled him to go on to the grid: all he needed was a road-racing motorbike.

Racing motorcycles cost. They cost a lot and Frank didn't even have a little. Approaching the old man on the subject would have been pushing his luck. Discretion here was the better part of valour, Frank felt; his parents would come to terms with the entire venture only after the fact, if at all. Letting him ride in enduros was one thing, but here he was talking about a real road race. The likely answer would be a firm no, followed (or maybe even preceded) by a firm slap.

To the rescue came Frank's cousin, with the suggestion that Frank might *borrow* a racing bike.

Borrow? Who would lend anybody a motorbike to race?

It was at this point that Frank first encountered the Holy Edge, the patron saint of all competitors, and not only in motorcycle racing. The Holy Edge that everybody was determined to have on his side, or at least never to give away. The edge that (everyone believed) often amounted to the sole difference between winners and also-rans; the edge without which a racer might as well stay home. The edge, which could be anything, from a single horsepower (plus) to a single ounce of unsprung weight (minus). The edge that the writer Jack London invoked almost a century ago in his famous story about a pugilist: "One Piece of Meat".

Because if you give away one piece of meat, you give away the edge, you give away everything. The fellow with the extra calories or the extra cc's will punch you out. You can be scratching your cases in the corners and the other guy will still pull you on the straightaway.

The edge in Czechoslovakia around 1950 translated into the latest CZ with an engine capacity of a hundred and fifty actual cubic centimetres for the lightweight (under 175 cc) class. Before the new machines came on the market people would happily contest the class with CZ 125s if they had nothing bigger. Now nobody would ride 125s anymore; what was the point? The old CZs were not competitive. They'd be giving away the edge.

The Holy Edge always works for the guy who has nothing, and so it worked for Frank. A year before he would have been without a ride. Now he had no trouble borrowing a leftover CZ 125 from the club. Let the kid have it, why not? It may be good for a laugh.

It was in the town of Znojmo that the flag dropped for Frank for the first time. A true road circuit, not man-made, not even man-adjusted, except for some hay-bales in front of the lamp posts. (Or sandbags, more likely, because you wouldn't waste good hay on racers when it could be fed to the cattle.) But a lot of people were standing on the curbs, leaning out of the windows, sitting in the trees, thousands of people, since road racing was as popular in Czechoslovakia as anywhere else in Europe. Znojmo was unlikely to offer a better entertainment for a Sunday afternoon.

Among the spectators were František and Marie Mrazek. Somebody had tattled. His parents had learned about Frank's road race the week before—and they did nothing to stop it. No words, no slaps. There was no telling why, but perhaps the old man wanted Frank to learn the hard way. It was the same way he had let Frank learn about the game of soccer a few years earlier, with people laughing their heads off at Frank's expense. There was no doubt about it, he did not have the old man's touch.

From the grid Frank couldn't see his father, but he caught a glimpse of his mother on a kitchen chair set on the curb next to the start/finish line. His borrowed two-piece leathers were riding up on his back. His borrowed helmet—also leather, for even Grand Prix riders wore nothing else back then—was sliding down on his nose. As the flag dropped, he could feel his borrowed kidney-belt snapping undone. Only his borrowed goggles were fine: they didn't even pinch.

He couldn't remember which row he was starting from but

it didn't seem to matter. He was sure he had stalled his engine when he released the clutch because he couldn't hear it anymore. It was lost in the scream of a hundred banshees rising from the grid. Then he thought that maybe he hadn't stalled it after all because he was being carried along by a herd of wild horses towards some unseen, mysterious point, probably the edge of a cliff. It must have been a cliff because the horse in front of him had suddenly stopped. The road tilted, rose, floated by overhead, then it disappeared behind him. He was facing a wall of dark suits and white faces. Where did all the motorbikes go?

Frank was sitting on his ass. Evidently he had made the short straight leading to the first corner. He had made it with no trouble, unless one counted a state of frozen panic. Then, as the rider ahead started braking, Frank had promptly rammed him from behind. Now he was sitting on the cobblestones, quite unhurt.

"I told you" was his mother's only comment, possibly for the benefit of onlookers, as Frank pushed the bike back to the start/finish line. She repeated it in private on the way home: "I told you, you should stay away from it."

Mrazek Senior said nothing. It may have been a good sign or a bad sign, but Frank didn't care. His tailbone was hurting now, every dull throb sending a message of grim, vengeful determination to his brain. This wasn't fucking soccer. He was going to show them.

He did show them, a few weeks later. The place was called Svitavy, somewhere in Moravia, though Frank hardly noticed where it was. The same kind of circuit, running through the same cobblestoned streets. The same high curbs, the same lamp posts, the same houses with grey plaster walls on both sides of every corner. And the same people, the crowds standing or sitting right at the edge of the road, though some of the smarter spectators would be crouching in the trees.

Frank was riding the same bike, a three-speed CZ 125, lent to him by Josef Hejmala, a racer who had graduated to a new 150 cc four-speed machine like most of the guys. Apparently a lot of local heroes were on the grid: well-known Moravian road racers, some with their own fans, camp followers, and pit popsies. And Frank, sitting among the heroes. Far back on the pillion of a 125, in the riding style of the period. With a total road racing experience of one short straight and a spill before his first corner.

Some rookies in his place—most, perhaps—would have been awed. They would have shown respect. Maybe they would have worked out a plan, some idea of trying to stay with a veteran through a few corners, learn his lines, pick up a pointer or two. But no such notion entered Frank's head. He had never really followed the sport. He was no groupie, no enthusiast. He realized that some of these guys were supposed to be big, but he couldn't even remember their names. In through one ear they went, and out through the other. He had no respect for anybody. Maybe because he didn't know enough, sure, but also because he didn't want to know. Instinctively. Then, or later.

His eyes were fixed on the flag.

This time he did not crash in the first corner. He crashed in the fourth corner, about halfway down the three-kilometre circuit. He picked up his bike, re-started, and chased furiously after the field. Two or three laps later he crashed again. He crashed in the same place. He picked up his bike and re-started. On the next lap, in the same corner, he crashed again.

It was worse this time because when he got to his feet he couldn't see the bike. For a split second he thought that he was going crazy. He kept turning around, feeling like an utter fool, until he caught sight of some onlookers pointing.

The CZ was lying on its side behind a line of spectators. Apparently the sturdy fans of Moravia had jumped to one side as the bike was sliding off the turn between their legs, then closed ranks again to continue watching the race. Frank fished out the machine from behind them and remounted. He had about eighteen laps left of a twenty-five-lap race.

He did not crash after that. He rode with gritted teeth, elbows tucked in, knees gripping the tank. Soon he started talking to himself, a habit he would have all his life. He spoke in muttered bursts of anger, like the frustrated father of a hopeless son. Come on, you asshole, get moving. Get on those guys' tails, stop screwing around. Go, clumsy son-of-a-bitch, go.

There was nothing ahead of him now but the ribbon of the street, snaking among grey houses. Then, suddenly, an obstacle he couldn't recognize. Was that a person, another rider? Bastard, why was he in the way? Goddamn it, get around him, go. Then another object, then another: trees moving on the road in the same direction, trees, bushes, or riders, Frank couldn't tell. He didn't care, either. Pass them. Inside, outside, in-between, go.

The first thing he could distinguish was a huge black-and-white pattern flapping in his face. It was just dropping in front of his eyes like a curtain. Jesus, the chequered flag. How's that possible in the middle of a race? When they still had about a hundred laps to go?

They had, in fact, no more laps to go. The race was over. And it was only when the starter offered him the chequered flag as he was coming around again on the cool-off lap, that Frank realized he had come first. He had won.

He had actually won the race. He had beat the big heroes on his 125! Even after three crashes, he had come first! And the most astounding, the most unbelievable part of it was that Frank didn't even feel surprised. He knew that he ought to be, but he wasn't. He went through the motions as expected: he tried to act surprised at first, later tried strutting a bit, still later pretended to be modest again. But deep down he felt that winning was only his due. He just couldn't help it, he was a winner. A champion like Frank Juhan. He had known it since the age of eleven.

He walked the track with his cousin after the race. Crashing three times in the same place was stupid, there had to be an explanation. And there was: a puddle with some fine sand in the cracks between the cobblestones, right on Frank's line, the very spot where he'd put his skinny front tire under braking.

He had been squeezing his front brake over that patch of wet sand. No wonder he'd lost traction—but why did he *stop* crashing after the third time? He never changed his line, as far as he could remember, he didn't even notice the puddle throughout the race.

But perhaps after the third crash that sandy spot was already past his braking point. Perhaps he was already on the gas, starting his drive out of the corner, unloading his front wheel. Perhaps he had started going *faster*, that's why he didn't crash again.

So there was a lesson: if there's a problem, go faster. Maybe you're in trouble only because you're too slow.

Frank didn't say this to his cousin. Maybe he didn't even say it to himself. It became a thing beyond expression or recall, something tucked away in his mind, something only his body or his hands would remember.

At home his father made no comment, but neither did he try to stop Frank from going to the next race. Frank went, and won again. He didn't even have a spill. Then, riding the same 125, he won another race. It was the last race of the season and his father was there to see him take the flag.

He didn't say it to Frank, but Jiří overheard him saying it to their mother when they got home after the race. "Maybe that kid has something" was what František Mrazek said to his wife. "Maybe I should give him a hand."

"You're the same crazy as he is" was Mrs. Mrazek's reply.

3

THE LINE

In the beginning there was the line.

Road-racing courses, whether they are ordinary roads marked off by hay-bales or specially constructed circuits (called "man-made" circuits to distinguish them from regular roads made, presumably, by the Tooth Fairy), consist of a series of points. Racetracks go from point A to B to C and so on. These points are only dots in space. They are like a child's puzzle. You need something to string them together before they can take on a shape.

The imaginary string that ties these dots together on a race track is called The Line.

The Line is the great secret, the eternal mystery, the supreme quest of every footpeg-scratching, knee-dragging acolyte. It's Merlin's spell, Aladdin's lamp. It's the alchemist's elixir that prolongs life and turns base metals into gold. To know The Line is to know everything.

Simply put, The Line is the racing line. It is the ideal way around a given racetrack. It is the path a rider must follow if he wishes to be swift and smooth. It is The Line that leads to quick lap-times and to the winner's circle. Ultimately it's The Line that leads to a sponsor, a works bike, a championship. It's the rail for the express train. It's the groove in which the Sacred Traction dwells. It's the yellow brick road leading to fame and fortune.

Where is this magic line?

One clue: the word is sometimes used in the plural, as in speaking about someone's "lines". To learn a champion's lines is to learn the computer code to his bank account (even though on the track he freely exhibits his lines in plain view of everyone). "Hey, let me see your lines" is a friendly request riders often make of each other at morning practice. Some complain that they "haven't found their lines yet" around this or that corner. Others like to show off their lines. In fact, Las Vegas showgirls couldn't look more eagerly for a chance to display their lines than some racers do.

Only it is not that simple.

Young peg-scratchers generally believe that The Line exists. Later some begin to suspect that it doesn't. Whatever they conclude, if they have looked for the line outside of themselves, they've been searching for it in the wrong place. Because The Line, like gravity or relativity, is and it isn't. It both does and doesn't exist in real time and space.

The Line is an idea. It is a perception, much like the molecules floating in space that take on various shapes as a result of changes in human perception. Under a microscope a galaxy of space rubble swims in the void, but what the naked eye sees is a toenail. The difference is that while anybody with normal eyesight can perceive a toenail, one needs a special kind of vision to perceive The Line.

Of course, a person could arbitrarily connect the dots that make up a racing circuit with any number of lines from start to finish. The problem is, most of these lines would not be "the" line. Nor would any one line be the *only* line (even if some racers believe so). If only one line existed track-owners would be able to paint it on the asphalt.

The ideal line around a racetrack is the line a rider must follow in order to reach the finish before anybody else. It may well be the shortest line, but whether it is or isn't, it must be the fastest line for *him*. It doesn't have to be the fastest line for everybody. If it's the fastest line for nobody, it could be the ideal line for one rider—ideal, that is, if he happens to be the fastest rider on the track.

What is the shortest line? Obviously, a straight line between two points. What is the fastest line? The answer to this question makes racing both a science and an art.

Few (sober) people have any trouble figuring out the shortest line between two points on a straight stretch of road. The shortest line around a curve is a little harder to figure out even on paper, and it's especially hard when one has to do it at speed by imagination or by memory.

A diagram gives a bird's eye view of the whole track, making it possible to perceive the shortest line. But crouching behind the fairing of a motorcycle at 100 mph all you see is that somewhere ahead there's a turn. You won't see if it's a sharp or a gentle turn. Depending on the topography you may not even see if it goes left or right (unless a sign tells you or you happen to remember it). Many riders will be glad if they can figure a way to go around it—any way, let alone the shortest.

The shortest line ought to be the fastest line provided everything else is equal. When, however, other things are not equal, it may not be. Remember, the winner of a race isn't the guy who clocks the shortest distance on the odometer but the one who gets to the chequered flag first.

There's a temptation to turn the equation around and say that the fastest line must be one along which you can drive at the greatest rate of speed. Since you attain the greatest speed on the straight, the fastest line must be the straightest. It's this (very logical) perception that leads many racers to try and "straighten out" a curve—that is, to increase its radius as much as they can by entering and exiting it at its widest possible point. On a motorcycle this technique is combined with a brief prayer that one can get around the turn before running out of ground clearance (or traction) in the middle, or out of pavement at the end of it.

The same desire for the fastest line can also lead to the opposite tactic of "squaring off" a turn: going in deep, braking hard, making a relatively short, tight semi-circle, then straightening up to start the so-called drive out of the turn as early as possible. The theory here is that since motorcycles must be going straight to go fast, one should spend as much time going straight as one can.

As Rudyard Kipling pointed out, there are nine-and-sixty ways of constructing tribal lays and every single one of them is right. (Or wrong, as the corollary seems to be, though Kipling makes no mention of this.)

An actual racetrack isn't lines drawn on a piece of paper. What if the shortest line is also the bumpiest? What if it has lots of fresh cement dust over an old patch of oil? Or worse, a fresh patch of oil over lots of old cement dust?

Or what would be The Line around a corner where someone beat you last week, say, by braking later and harder? The guy who beat you has all the motor and sticky rubber in the world. He has the horsepower and the traction for an early drive out of a tighter turn—which you lack—but if you let off the brakes early you may retain enough momentum to go around a wider arc faster than he does. And should *you* beat *him* this week, are you going to call the ideal line around this corner "tight" or "wide"?

Or take the next corner. As everybody knows, there's only one quick line around it—and at present it happens to be occupied by the slowest competitor on the track. Are you going to queue up behind him or pass him on a "slower" line?

The permutations are endless. They are also meaningless, because the line, the ideal line, The Line, is not on the racetrack. The Line exists, all right—but it exists in a few human beings' minds.

In the 1980s it has existed in the minds of such world championship riders as Eddie Lawson, Freddie Spencer, Kenny Roberts, Marco Lucchinelli, Barry Sheene, Randy

Mamola, Ron Haslam, Wayne Gardner, Anton Mang, Wayne Rainey, Graeme Crosby, Kork Ballington, Jon Ekerold, Gregg Hansford, Mike Baldwin, Wes Cooley, Christian Sarron, Boet van Dulman, Jack Middelburg, Takazumi Katayama, Franco Uncini, or Ricardo Tormo. In preceding decades it has existed in the minds of Stanley Woods, Geoff Duke, Mike Hailwood, Giacomo Agostini, John Surtees, Bob McIntyre, Hugh Anderson, Paul Smart, John Cooper, John Hartle, Derek Minter, Gary Hocking, Steve Baker, Cal Rayborn, Kel Carruthers, Dick Mann, Phil Read, Bill Ivy, Jarno Saarinen, or Tarquinio Provini. This is just a sample because the list, though not endless, is long. Currently it includes such national championship riders as Kevin Schwantz, Doug Polen, and Bubba Shobert (U.S.) or Gary Goodfellow, Reuben McMurter, and Michel Mercier (Canada). It includes such masters of classic/modern technology as Battle-of-the-Twins riders Roger Marshall, Doug Brauneck, Dale Quarterly, Paul Lewis, John Long, James Adamo, or Stefano Caracchi. It probably includes about 300 to 400 top riders of the approximately 22,000 men and women who race motorcycles in North America, Australia, Asia, and Europe.

Only about a score of them will be household names like "King Kenny" Roberts; many of them may not be known even to aficionados or sports writers, except locally. They may be grizzled privateers; grass-roots heroes, who for one reason or another—money, family, luck—never made the national scene. They may be rookies just catching their first chequered flags in club races. But these riders are unlikely to search for The Line on the racetrack. The Line will sit in their heads behind their eyes.

In music it would be called perfect pitch.

None of this means that such people cannot profit by training. Everybody can profit by training; Mozart profited by training. It doesn't mean that they never give racing techniques a thought: they may (to the despair of their families or acquaintances) think or talk about little else. It certainly doesn't mean that they don't learn through experience. On the contrary, they absorb experience like a sponge, and learn something new every time they go around the track. They keep learning until they hang up their leathers.

It only means that nature has equipped this handful of people with an ability to take on, or fall into, a riding rhythm sometimes called "having the line" or "hitting the groove". It is a rhythm others feel only rarely, or only for short sections of the track. For gifted racers it must be like what we imagine soaring to be for hawks. For them it usually encompasses the entire circuit, lap after lap.

Having the line is a beat. It exists both in time and in space. It means being in the right position, at the right time, at the right angle, in the right gear. It is just as fluid, just as continuous, just as inevitable as a musical pattern. It makes everything fall into place. It co-ordinates a racer's mind with his body, his body with his bike, his bike with the track. Once you catch the groove it is almost impossible to consciously move out of it, just as it would be impossible to dance against a beat. The line is a harmony between track and rider, a movement to a music unheard but reverberating in every neural cell, a transfiguration, a trance, a ballet at speed.

And yes, of course, a *danse macabre*. It is waltzing with Death.

To say that danger or death adds a "spice" to such sports as speed contests is trivial. Death is hardly a "spice". It is the ultimate experience and common destiny of all living beings. To be in the shadow of death is to come as close as anyone alive can to the absolute unknown of the Beginning and the End. Death, whatever it may be, is the closing chord, the perfect balance, the quiescence of all scales, the final equilibrium of nature. Between it and life there is a short gap, maybe a foot or an inch wide, ending at the red-and-white speedbumps of the track.

This is where The Line runs, at the edge of this final mystery. It is not the "spice" of motorcycle racing: it is its essence.

Racers are reluctant to say much about this, even among themselves. They're almost certain to clam up or change the subject when outsiders raise it. Sometimes a few try to mask their reluctance with a false swagger, but more often they just shrug and keep their peace.

People who race are puzzled when outsiders call them crazy. They truly don't think that they are crazy, or have anything remotely resembling a death wish, but they know that this is a common view taken of their passion. Being as vulnerable to social opinion as all human beings are, some deny even to themselves that death plays any part in their fascination. They're having enough trouble with the outside world as it is. Or even with their own self-perception. It's easier for them to blank out the subject from their minds.

The accepted phrases among racers speak only of "the high" or maybe of "that feeling". It's an acute sensation, though by no means the sole reason for racing. As in other sports, people want to compete. They go on the track to dice with each other, to try and smoke somebody, to chase the quick guys, to achieve a personal best—if possible, perhaps to win. Many are fascinated with engines, mechanics, with the physics or aesthetics of automotive art.

But people on the track know that "catching the groove" is "a real buzz" no matter where they happen to finish in a race. If two racers are on friendly terms (and sometimes even when they're not) they'll grin at each other as they're pulling off the track after a great dice. They'll nod or raise a thumb. It signifies a shared and almost incommunicable experience. "Yes, I was out there, nipping at the edge, and you were staying with me. I saw IT—and I can see that you've seen it too."

Calling it a "death wish" is trying to fit something large into a small box. It's an outsider's attempt to reduce an experience beyond comprehension to a reassuringly familiar label. Also a label that serves to affirm the observer's own normalcy. *That* person has a death wish; *I* am in great mental health.

A racer no more "wishes" to die than a mountaineer wishes to fall off a cliff or a diver wishes to drown. A racer tries to avoid crashing as much as possible, with all the natural instincts of any living thing trying to avoid injury or pain or extinction. As much as possible—short of staying home. What he looks for on the racetrack is "that feeling". A feeling that, unfortunately, is not aroused by playing computer games. A feeling that is available only in the shadow of death.

In many ways this "high" is like other experiences of ecstasy. Having tasted it, people often want to taste it again. Perhaps it's reminiscent of a chemical euphoria because, in fact, it is. The changes that overdosing on adrenalin or other hormones aroused by the proximity of death produce in the brain are chemical changes. Their effect mimics a physical addiction. In some racers, at least, the desire for them resembles a bodily craving.

Other sports can satisfy competitive urges, animal spirits, delight in co-ordination or skill, aesthetics, friendship, self-mastery and the rest (and racers often dabble in other sports from downhill skiing to Ping Pong) but very few sports can produce that feeling. Sports that do produce it are either speed contests of various types—speedboats, bobsleds, pylon-racing with airplanes—or perhaps mountaineering and sky-diving. They all share proximate death as their common denominator.

The object is never death; it is only the price of admission. For those who would dice with God the stakes are high. The feeling doesn't come except in the presence of mortal danger. Tennis is a great game, but it doesn't produce that feeling. It is not aroused by synchronized swimming.

Yet merely scaring oneself doesn't seem to produce it either.

One couldn't find that feeling in the barrel of a gun, playing Russian roulette. At least, few if any racers are "thrill seekers" in such vulgar or pathological ways. Most would consider it a sacrilege to look for shortcuts to the supreme experience of The Feeling by simply making a game of it.

They're likely to view with cold disdain anyone who takes juvenile chances with death in any context off the racetrack. They look down on people who play "chicken". Some racers become almost priggish—whether in compensation or perhaps as a form of reverse snobbery—and tend to be more cautious in their daily lives than the stolidest briefcase-clutching citizen who has ever mounted a commuter train. They may never ride a motorbike except on the racetrack, saying that it's "too dangerous" to ride it in the street. Many racers dawdle in their family cars like old ladies. They use the scathing term "street squid" for anyone who takes chances on a public road.

There are exceptions—there are always exceptions—but as a rule the progression is this: As youngsters, future racers may behave like cheap thrill-seekers. They may go berserk the first time they sit on a motorbike; charge about heedlessly, crash, break bones. Soon, however—usually by their first or second season on the track—they grow from frenzied worshippers into solemn high priests of speed.

Before a race they carefully lay out the vessels of the Mass. They sit on the starting grid calmly, hands folded, waiting for the celebration to begin. When the one-minute board turns sideways, the switch goes behind the starter's back, the flag is unfolded, they lean forward. Not in excitement, only to put some weight on their front wheels.

They wait for the word, the blessing, the benediction. With grave introspection they raise their eyes to the flag. The flag that unleashes the spirit, the trance of the dance, the dance of death, the euphoric quest for the profound mystery of The Line.

4

The Whispering Years

There are racers who go to the track mainly to tinker with machinery. At heart they're really designers, fabricators, or engine tuners. By training they're often mechanics. After a few seasons of racing they usually end up wrenching for others.

Then there are those who tinker only because if they want to race, they must tinker. They have no other choice. If they had a choice, they wouldn't kick a tire. As far as they are concerned, driving machines is the goods and fixing them is the tax. (Some racers are ostentatious about their mechanical ignorance, like the great Belgian Grand Prix driver Olivier Gendebien, who once pretended to pour a can of water into the "radiator" of his air-cooled Porsche for the amusement of international moto-journalists.)

Without being ostentatious about it, Frank belonged to Gendebien's school of technical nonchalance. Maybe not quite or not completely. He was genuinely fascinated, even enchanted, by automotive hardware, down to its very sound and smell. The sight of a supercharged DKW could send him into a trance. Still, Frank was no grease monkey. He was perfectly content to sleep in his bed the night before a race while his father and brother Jiří worked on his CZ in the garage. For Frank, gutting or building motors was like shaving and getting dressed. You've got to shave and dress if you want to go to the ball—but the ball itself begins at the racetrack.

It was for this reason that Frank and his father made a good team.

Old man Mrazek liked tinkering. He also liked wheeling and dealing, for which Frank had no head at all. Actually his *head* wasn't the problem, it was no worse than most people's, but deal-making could never engage his full attention. Nice bike, you want a zillion for it? Fair enough, here's your money, let's start it up.

Not while the old man was around, though. If there were shots to be called—important shots, money-shots—he'd call them. Frank could fetch and carry; he could draw some water and hew some wood, then get on his bike and race. That was the deal.

A fair deal, at least in the beginning. It suited Frank to the tip of his laced leather boots. His attention span in those days (it was to improve with age, like everyone's) was about as long as a rhesus monkey's. Things that didn't involve racing could engage it for a maximum of ten seconds, maybe. Racing was life. Everything else was only a vague background buzz, sometimes pleasant, sometimes irritating, but never important. A person has only so much attention. Why spend it on things that weren't going to improve your lap-times?

Racing was life, the rest was "politics". For Frank this word encompassed everything, not only from international affairs to office intrigues (which it does for a lot of people), but right down to quarrels with friends or family. A fight with your girlfriend was politics. Any conflict of issue or personality, wherever it occurred and for whatever reason, came under the heading of politics for Frank. And if it was politics, it was bad. It was just about the worst thing on earth. Politics was exactly what the wrong type of human beings played at non-stop, the ones who were screwing up the world for the right type of human beings who wanted to do nothing but keep their mouths shut and race. For Frank this pretty much summed up the history of the civilized world.

For someone with such an outlook, any society in any period might appear less than ideal, but Czechoslovakia in the early 1950s was rather less ideal than most.

For instance, right at the start there was a matter involving a really fast 350 cc Jawa. A racer from the Slovakian city of Bratislava, a fellow named Jura Ejem, was cleaning up on this great twin-cylinder two-stroke: he was just running away with it. And the word at the end of the season was that Jura might be ready to part with this awesome machine for 20,000 crowns.

A 350! Frank had always dreamed of big bikes. From where he was sitting at the time—the saddle of a CZ 125—a 350 Jawa seemed to be the biggest thing on earth. A mastodon might have been bigger, but there were no mastodons in Bratislava. Frank simply had to have the Jawa, and old man Mrazek did not appear to consider the obstacle of 20,000 crowns altogether insurmountable.

It was at this point that politics interfered. It was honest-to-goodness politics—something that had to do with the country's economy, at least as it appeared to the comrades newly occupying old Hradčany Castle at the people's democratic capital of Prague. The political types—the bad human beings who are forever plotting the downfall of good human beings who are only trying to race—decided at this precious juncture in Frank's life to revalue the nation's currency.

Twenty thousand old capitalist crowns would have been a manageable sum. Twenty-thousand new people's democratic crowns catapulted the Jawa's price out of Frank's reach, or just about. When one added the personality of Daddy Mrazek to the mix, the whole dream seemed to recede into infinity faster than Einstein's universe.

"Starý" Franta (as most people referred to old man Frank to distinguish him from his son, "Mladý" Franta or young Frank) wasn't given to impulsive gestures. His nickname in

Brno—not a town noted for its philo-semitism, one might add, but then few places in this world are—was "white Jew".

The amazing thing was that such a man, who with a thousand crowns in his pocket would rather walk eight kilometres from the central railway station in Brno to his suburban house than pay four crowns for a ticket on the bus, did come up with the 20,000 new crowns for the Jawa. As Frank would put it in his own brand of English many years later, his father somehow "squeezed money" together. He actually purchased a 350 cc two-stroke racing twin for his fifteen-year-old son.

Now Frank had the Jawa, and he soon became king in the middleweight classes. The bike was everything he expected it to be, and then some, except for its stock ignition. The solution was to adapt a Bosch magneto—everybody did that—which stuck out about a mile from the crankcase. It wasn't just a problem of aesthetics. In right-handers the thing would touch down, and then whoosh! The magneto would keep flying off the end of the crank, literally, and Frank would be searching for it on the track. He would be on his hands and knees looking for the damn magneto in the middle of a race, to the mild amusement—not so much of the spectators who were used to mechanical reliability, East European style—but of visiting western competitors.

Not that Czech technical ideas were bad. On the contrary, many of their original inventions or innovations would be picked up thirty years later by the Japanese for their miraculous rice-burners, meticulously refined and flawlessly executed, of course. Take reed valves, for instance. They had reed valves in Brno on the CZs in the 1950s. What East Europeans did not have were decent tools, machinery, and materials.

Not even proper rubber. Everybody was running street tires, hard as rock, narrow as a wasp's waist, slippery as glare ice. The only advantage was that, unlike soft, sticky racing skins, a set of rock-hard tires might easily last a season or two. You couldn't shred them with dynamite.

Another advantage of Brno was the circuit itself, the Masarykův Okruh, or "Masec", as the locals called it. It was a great place for the young set to mix and match after the races, especially in a town of Victorian socialism where you couldn't take your girlfriend to a motel. Named after Tomáš Masaryk, Czechoslovakia's first president, the Masec was a real international track, part of the World Championship Grand Prix circuit, with great foreign teams coming to race on it at least once or twice a year.

The Mitters, for instance. Frank was walking with his father through the pit area in Brno one day when he saw, right in front of his nose, on the other side of the rope separating the visiting teams' pits from the local racers', a young boy working on a racing bike. He was talking with an older man in German as he worked, a man who looked like the boy's father. The two of them could have been Frank himself with old man Mrazek tuning their rickety CZ. It was a spitting image, and they were so close that Frank could have touched both men with his extended arm. Only they might as well have been on the dark side of the moon.

The bike on which the boy was working was an MV Agusta. A 125 MV Agusta single. An actual MV Agusta thumper, a red-hot one-lunger like world champion Carlo Ubialli's. It was something to die for, or maybe to go quietly insane about, and there was this boy who could have been Frank himself casually trying to fix the thing. Frank simply couldn't believe his eyes.

He said as much to his father, in Czech, of course, and suddenly the older man turned, smiled, and started talking to them in Czech. He spoke the language fluently, with only a slight accent, which wasn't suprising because he turned out to be a former ethnic German from the Czech Sudetenland: that distant part of the world British Prime Minister Neville Chamberlain declined to go to war for when Hitler slashed it off the young Czechoslovakian Republic's body in 1938. Of course in 1945 it became the Czechs' turn to expel most ethnic Germans from the same Sudetenland now re-annexed from Hitler's defeated Third Reich—expel them, even though they had lived there for generations and Hitler hadn't exactly consulted them about his plans for a greater Germany seven years earlier. It was Politics again with a big P, though Frank knew little about it, and cared even less.

The Mitters had been expelled along with the rest in 1945. They were forced to leave everything behind and move to Stuttgart in West Germany. A bitter hardship at the time, which turned out to be a huge favour three years later when all the blessings of a new People's Democracy descended on Czechoslovakia. Frank remembered Mitter's wry remark, even though political chatter usually put him to sleep. Looking at young Gerhard Mitter's sparkling MV Agusta, he could also believe it. Frank wasn't unhappy in socialist Brno—he never gave it a thought one way or another—but for such a bike he would let himself be expelled any day.

This all came to light when the elder Mitter asked for some assistance with his son's motor—it appeared to have dropped a valve—and Frank's father was only too glad to oblige. One of Starý Franta's trucks dispatched the MV to the Mrazeks' garage where it was fixed in time for the race, free of charge, of course. Frank was actually tempted to charge admission to his friends, the young peg-scratchers of Brno, who all came to gape at the MV while it was being repaired. They just stood in a semi-circle, like a pack of wolves, dripping saliva. It was the greatest toy any of them had ever seen.

After that Frank and young Gerhard became friends, as did the two fathers, and it became possible to acquire the odd piece of western equipment through the Mitters, things that with a little fiddling could be fitted to Frank's machines. Like some Avon tires. Like a Bosch battery and coils to run the Jawa with total-loss ignition: a great improvement on searching for broken-off magnetos in the middle of the racetrack. It was with such bits and pieces from visiting competitors that all Czech racers kept going, including even the big guys, the works riders of the Jawa and CZ factory teams.

Another racer stood out among the visitors: Mike Duff, the fastest Canadian, a world championship rider. Frank wasn't intimidated by Duff any more than he had been by the heroes of Moravia the year before—in his own mind Frank continued to be the Frank Juhan of childhood memory, the champion who did not have to respect anyone—but he couldn't hope to beat world championship riders on his clunker. So he was content to keep Duff's tailpipe in sight for a corner or two, then watch him in his father's garage when Duff required some technical assistance. Old man Mrazek was happy to help out visiting riders, freely though not disinterestedly, because there would always be some tit for tat, some exotic western part or tool.

The amazing thing was that Frank could even talk with Duff, become friends with him after a fashion, even though the Canadian had no Czech and Frank had no English. They both dipped into German, a language of which Duff had little and Frank even less. But bench-racing was bench-racing in any language and people who lived on the track understood one another. In a way it was easier to have a good chat with a foreign racer than with a non-racing fellow citizen from across the street.

Altogether it was a good season, that second one: racing the big Jawa, racing a new CZ 150 his father and some friends had bolted together in the garage, racing every weekend, finishing in thc top five most of the time, and even winning a few. It was a good season, and one night it abruptly came to an end.

It came to an end all at once, unexpectedly, in the manner of most catastrophes. The whole thing came out of left field for Frank, though he couldn't say that it came out of the blue because it happened in the dead of the night.

There was a banging on the door, then suddenly some men in leather coats were going through the house. It was all a blur and a few minutes later it was over. The men had gone through the house and then left, taking Starý Franta with them. The men were the secret police and they had picked up old man Mrazek. They picked him up and carted him off to the mines, for being a "kulak".

It was politics, of course. Some neighbour, some envious

soul, had done his patriotic duty. He or she had called the attention of the authorities to the fact that in their midst there lived an enemy of the people.

In Stalin's waning years—though Frank was barely aware of it at the time—the embryonic Czech communist state was faithfully re-living the earlier years of its Soviet ancestor, like ontogeny recapitulates philogeny in nature. For Frank, it only meant something stupid: people always talking in whispers for some reason.

During those whispering years there were a few intense, official campaigns. Entire groups were deported, exiled, or put into labour camps—people labelled "kulak" or "bourgeois" or "clerical-fascist", not to forget some "old-fashioned intellectual" or "class-alien" elements. At other times it was just individuals being denounced by friends or neighbours. If a hated person could be made out to belong to some high-profile "enemy" class, some group it was trendy to persecute just then, it was almost guaranteed to spur the authorities into acting on a denunciation. But even this wasn't necessary. If anyone, including a manual worker or a peasant, was accused of "slandering the regime" or "spreading false rumours", or maybe of being a "Tito-ist", a "reactionary", a "hoarder of scarce goods" or just plain "hankering after the good old days", he or she might speedily join the kulaks or clerical-fascists in the camps.

The list of possible accusations was almost endless. An alleged crime against the people did not have to amount to a breach of some law; it did not even have to be an act. An attitude could be a crime. Nor would it have to be proven. Indeed, often there would be no occasion to prove or to disprove it, because there would be no trial. Not even a mock trial, not even a Star Chamber procedure of some sort. There would simply be a knock on the door, followed by a blur of leather-coated men in the night.

In the case of Starý Franta there was an additional twist, a sort of insult added to injury, because it turned out that the anonymous denunciation against him alleged that he had been drinking, playing cards, going to restaurants, living in the fast lane. In the days of Stalin's Victorian socialism these were regarded as serious, even provocative misdemeanours, although playing cards or dining out was not against any law. But what it was supposed to show was a bourgeois attitude: perhaps even a hankering after the good old days.

Fine; except old man Mrazek never drank. He took a dimmer view of alcohol and nicotine than most people take of arsenic. He would not dream of darkening the door of a restaurant—what, spending good money on a bad meal!—and as for playing cards, he would just as soon have put a match to his hard-earned crowns as to have bet them on a pair of aces. Gambling would have made no more sense to him than hitting his own hand with a hammer. Living in the fast lane? The old man wouldn't have known where the fast lane *was* in Brno, and would have thought any man a fool who did. It was like accusing the Pope of consorting with abortionists.

Starý Franta would have said all this to his accusers or his judges if they had taken him to some place where he could say it, but as there was never any trial he said nothing. There was certainly no point in saying anything to his Russian bosses who ran the uranium mines near the city of Příbram in Bohemia. The Russians wouldn't have cared; but luckily what they did care about was something infinitely better.

The Russians, thank God, cared about uranium. There they were, near godforsaken Příbram (where most of them wanted to be as little as most Czechs wanted to have them there), trying to meet their production quotas of precious, hard-to-mine uranium, with little help except from some ancient machines and a bunch of imprisoned kulaks and hankerers after the good old days.

For the Russians, meeting their production quotas meant a speedy return to the civilized environs of Leningrad or Moscow, with maybe a dacha and a chauffeur-driven Zim or Pobeda thrown into the bargain. Not meeting their production quotas, on the other hand, could mean anything, including an accusation of sabotage, Tito-ism, pushing the wagon of the imperialists or hankering after the good old days. It could mean—Stalin was very much alive then for the Russians, too—an eight- to twenty-five-year stint on one of the islands of the Gulag archipelago.

The ancient machines at the mines didn't work, or worked intermittently at best. The Russians couldn't fix them. The clerical-fascist and old-fashioned intellectual prisoners couldn't fix them. But František Mrazek could, and he did. He could fix anything, as long as it was a machine. If it could be chiselled or welded or soldered or hammered into shape the old man could do it. Soon after he was taken to the mines the ancient machines started clattering again. The uranium quotas were being met.

None of this was as simple or painless as the telling of it makes it sound later, but the final result was that Starý Franta showed up—actually showed up, in the flesh—at the house in Brno after only about eight months of having been taken away from it by the grim leather-coated men. Showed up on a pass every other weekend. Showed up, while still a prisoner in the uranium mines. And not only showed up, but spread out on the kitchen table slabs of jellied marmalade, tins of oily sardines, and even jars of black Russian caviar for his wife and his children.

The food and the furlough were down-payment from the bosses against their own future dachas and Zims back in Moscow, a future held in old man Mrazek's tinkering hands. They knew when to use a stick and when to use a carrot, those Russians. You can lead a horse to water but you can't make him drink; you can keep a man in prison, you can beat him, starve him, or even shoot him, but you can't force him to fix machines. What were a few tins of fish or a few grams of caviar when set against the holy quotas that were being met?

As to a prisoner escaping, there was never any danger of that. Where all borders are sealed, where the entire country is more or less a prison, where would a prisoner escape to? Where would he find a job, or even a place in which to sleep, without the proper papers? And what would happen to his wife and children? Why would he *want* to escape, anyway, when his family was being treated to sardines? A treat more valuable than gold, a treat that could be traded for anything from eggs to shoes. Exotic tinned sardines that few families had even laid eyes on since before the war (or since those good old days that were a crime for people to hanker after).

It was sheer luck that old man Mrazek had clever hands, because without them the family would have starved, or God alone knows what would have happened to them. The leather-coated men had (naturally) confiscated the property, the trucks, the horses, the wagons, after they had taken away Starý Franta, leaving nothing but the house itself. For six months Frank's mother couldn't get any answers from anyone. She was just running around in circles, the officials wouldn't even talk to her, neither at the local council nor at the police. "Well, if they arrested your husband, what do you want from us? They must have had a reason. But what about your big house, now that there's only you and the children? Perhaps we'll requisition part of it, there are lots of poor families in Brno who need housing. Especially if you keep pestering us about your husband" was what one high official said to Frank's mother when she got in to see him after a month or two.

All this changed when the old man became a friend of the Russians. Everything was still gone, of course, including the head of the household, but at least there was plenty of food and nobody was threatening to take the roof from over their heads. The Mrazeks could live and Frank could continue going to school—though, as he was to discover soon, he was still a kulak's son, a bad seed, a suspect name on the records with a red circle around it that couldn't be scratched out in a lifetime. Not even when his father came back from prison, released for good, after about two years in the uranium mines.

Old man Mrazek was released because no less a personage than Comrade Klement Gottwald himself, in those days president of the Socialist Republic of Czechoslovakia, wanted him released. Or perhaps it was only some lesser officials who figured that Comrade Gottwald would want to release the man with whom he had had a friendly chat in his office. It all sounded like Alice in Wonderland, and in a sense it was—but in another sense it was perfectly simple.

The uranium mines were important. They produced some vital stuff for the nuclear industry of the entire Soviet bloc. In addition to prisoners, they also employed some civilian workers.

That year the best civilians in the mines, the good Czech proletarians, or Stakhanovites, as they were called after a famous Soviet "hero of labour", were invited to take tea with Comrade Gottwald in his office. This was a great honour, and the civilian workers said yes, sure, with bells on. But wait a minute: only if František Mrazek comes with us. He's the chap who's making the machines go.

The communist officials went into a bit of a tailspin at this. "But he's not like you," they argued. "He's not a good proletarian but a kulak and a prisoner."

"Be that as it may," replied the smug Czech heroes of labour, "if he isn't coming with us, we're not going either."

It could have been a scandal. It would have made waves if the best workers had refused to see Gottwald. It might have displeased the Russian bosses. So old man Mrazek went with the rest of the Stakhanovites, drank his tea, chatted pleasantly with the great Comrade President, and was released a few days later.

Why not? After all, going to restaurants and playing cards was no crime, and if Comrade Gottwald (a man who had sent some of his oldest friends and comrades to their deaths) didn't mind that old Franta was a rotten capitalist, why should the lesser comrades care? There would be plenty of kulaks left in the mines.

Apparently the officials' idea was that Mrazek would stay on as a civilian worker and keep the machines going, but Starý Franta was a stubborn man and he refused. As far as he was concerned he had nothing more to lose. The officials had taken his land, his horses, his trucks, and now if they liked they could take him too. They could arrest him again, they knew his address, he wasn't going anywhere. But he wasn't goddamn volunteering for them. He wasn't going to be separated from his family for another two or five or ten years. He wasn't going to go back to the uranium mines, not for the Russians, not for a few tins of sardines, not for cups of tea with the great Comrade Gottwald. Not of his own free will, anyway.

Who could read the officials' minds? For some reason, they let the matter drop.

It wouldn't be accurate to say that Frank was not happy to see the return of his father. He was happy enough—but what he was ecstatic about was the return of his mechanic. Frank, no less than the Russians, needed someone to keep the machinery going. The two years while Starý Franta was away—almost two entire racing seasons—were little more than a waste. Frank was a good boy and a loving son, but he had the normal selfishness of youth coupled with the special selfishness of champions, poets, and other stubborn souls. Politics were interfering with his track time.

He did make the grid once or twice while the old man was away in the mines, but it wasn't the real thing. Still, he raced, and he did well enough for some people to notice. The one person who noticed in particular was a certain Captain Mošna. If a racer was to be noticed by only one person in Frank's neck of the woods, no one could have been better. Captain Mošna was the manager of the all-Czech racing team for DUKLA-Prague.

DUKLA was great. DUKLA was even better than SVAZARM. While the paramilitary sports club of SVAZARM embraced motorcycle racing for civilians, DUKLA embraced it for the army itself. It was the armed forces' club, and in Brno it was run by a reasonable sort of chap named Captain Knapp. In smaller, regional commands a civilian might make a guest appearance in the army club's colours, which was how Frank got to race for Captain Knapp's team in DUKLA-Brno. The big time, of course, was DUKLA-Prague, but to make Captain Mošna's team in the distant capital a competitor had to be a regular army man.

And Captain Mošna had noticed. At least, that's what he had said to Captain Knapp, and that's what Captain Knapp had said to Frank. By then Captain Knapp had become a friend, sort of, and he wouldn't have said it if it hadn't been true.

He even gave Frank a phone number for Captain Mošna in Prague.

It was an interesting situation. In 1954 Frank was seventeen years old. He had applied to technical college, but as a kulak's son there was little hope for acceptance; he wasn't even surprised when the school turned him down. Now his capitalist father was back in Brno, but what was Frank going to do with his life?

No one in his right mind was keen to join the army in Czechoslovakia in those days, but when a boy turned eighteen the army just took him, like it or not. Unless he was a college student or, God forbid, a cripple. Frank wasn't either, so sooner or later he would be marched out of the induction centre clutching his kit-bag, shaved as bald as a turkey. But if he joined voluntarily, a year ahead of his time, he could pick up that telephone right away and talk to Captain Mošna.

Then he would be a regular army man himself. He'd be eligible. If Captain Mošna wanted him, after his basic training he could be assigned to DUKLA-Prague. As a certified good guy, a sports officer in the People's Red Army. Rank: Second Lieutenant. Duties: racing.

Any fool could see that racing motorbikes as a second lieutenant in the People's Army was better than working as a mechanic's apprentice in some factory, which would have been Frank's only other option. Not even a real choice, because from the factory he'd be called up after about twelve months anyway. As a kulak's son, without anybody certifying him as anything.

So Frank volunteered with no hesitation. He even smiled when the army barber shaved his skull. He smiled when he was marched to the barracks with the other recruits and shown his bunk. He only stopped smiling after a brief talk with Captain Mošna on the telephone.

There was a problem, Captain Mošna said to him on the phone, heaving a deep sigh as he said it. Yes, the captain had given his number to friendly Captain Knapp to give to Frank. Yes, they had been talking about Frank racing for the national army team. But there was still a problem. Frank should go through basic training, then call him back. Maybe by then the problem would sort itself out.

The next two months were not happy ones. Basic training is no picnic in any army on earth and the Czechoslovakian People's Army was no exception, to put it mildly. Cold, shorn, grubby, sleepy, and miserable, Frank was going through boot camp. He was running, jumping, marching, saluting, scrubbing mess halls, oiling weapons, wading through mud, polishing brass, drinking thin soup, and listening to political lectures with the rest of the bald-headed recruits. Unlike the others, he was also chewing his fingernails. He was casting sidelong glances at the black telephone affixed to the wall in the staff sergeant's shack.

After eight weeks, the recruits were sworn in. They stood at attention listening to a wooden-voiced colonel outline the course of their lives for the next two years. Then, while the others scrambled for their first forty-eight-hour passes, Frank made straight for the telephone.

The line to Prague came through in a surprisingly short time. It was a good connection. Captain Mošna's voice sounded loud, clear, maybe even lively. The only sad thing about it was his message. The captain didn't want to say what it was over the phone at all. Even when Frank picked up his forty-eight-hour pass and took the bus to Prague, he would only whisper it, having first glanced carefully in every direction.

The problem hadn't sorted itself out, and it was not likely to do so in the foreseeable future. Frank, unfortunately, was the son of a kulak, the scion of a capitalist ex-owner of several trucks and horses. Such people didn't make good second lieutenants in the People's Army. This being so, it made no difference what kind of racers they would make.

Captain Mošna regretted if he had given Frank the wrong impression, but what the hell. As Frank could appreciate, people couldn't tell that he was a kulak's son just by looking at him on the racetrack. Politics, though important, were not visible at the start/finish line. Now that they had become only too painfully visible, Frank would have to stay a private. Kulaks' sons might not make good lieutenants, but they made fine privates in the People's Army. It wouldn't be such a bad life, Captain Mošna added. Frank would just scrub a few hundred floors, oil a few hundred weapons, and wade through a few hundred patches of mud for the next couple of years.

5

THE GOOD SOLDIER

And, of course, he wouldn't be racing motorcycles.

It seemed to Frank that on this single point everybody was in agreement. The whole world, beginning with his own parents, then continuing with the entire incomprehensible mishmash of the political and military universe. No one appeared to have any purpose in life other than to prevent him from racing. God alone knew why it was so important to the world that Frank shouldn't race, but there it was. A cordial accord between all people, from such a harmless soul as the balding lieutenant-general in Prague to such a complete shit as Captain Konoupek.

The sequence of events saw Frank assigned to the central military barracks in Prague after basic training. The compound lay behind the monument of a triumphant T-34, a green Russian tank sitting on its pedestal in front of the main gates. Prague was a cushy, enviable posting for a dogfaced private. No mud, no digging, no scrubbing. It wasn't just a stroke of luck or a result of somebody pulling strings. It was the logical outcome of Frank's special skill. He had a driver's licence.

Not that a driver's licence was such a rare qualification in Eastern Europe by the mid 1950s; it's just that it wasn't as common as it would have been in North America or in Western Europe. In America nearly every person of military age was a licensed driver; in West Germany, probably every third; in Czechoslovakia, every seventh or eighth. In the U.S. army a soldier who couldn't drive probably couldn't tie his own shoelaces. In the Czechoslovakian army during the same period—well, it was safe to say that a conscript who had a driver's licence was unlikely to be put on latrine detail for long. He had a fair shot at a posting as a military chauffeur.

Which is what Frank became. He drove trucks at first, but soon he was transferred to a nice grey Škoda with a balding lieutenant-general in the back seat. A big shot, but a pleasant, harmless sort of fellow who just wanted to inspect things. Frank certainly didn't mind driving him hither and yon all winter letting him inspect to his heart's content.

They didn't have much to talk about. Frank would try a sympathetic grunt or two whenever the lieutenant-general started mumbling about damned fools who were clueless about supply lists or logistics, but since he would lose the drift of the general's complaints within a few seconds he had to be careful not to grunt in the wrong places. As for making observations of his own, it was difficult. If the general had any buttons, Frank never discovered how to push them. The big shot displayed no interest in anything. He didn't follow soccer on the radio or read the sports pages in the newspapers. If he had a hobby other than inspecting things, Frank remained at a loss as to what it might be.

In spring the racing season began. Frank's duties were not at all onerous, as it happened; after driving the general home from his office on Saturdays he usually had nothing to do but sit in the barracks until Monday morning. One day he took a deep breath and asked the balding staff officer if he would permit him to go racing the following weekend.

"What?" The general sat up, evidently taken aback.

Frank repeated his question.

"You request permission to drive your motorcycle on Sunday?" asked the general. "Drive it in a circle for no military or civilian purpose?"

Frank allowed that this was so. It was undoubtedly one possible view a person could take of racing.

"No, no, no, no," the general said. "I don't think it's a good idea. It's an unnecessary risk. Remember, in the army we don't take unnecessary risks. First rule of warfare. Would you like me," he continued in a kindly voice, "to lend you some books on the subject?"

"Yes, please, Comrade Lieutenant-General," Frank replied, giving up on any further discussion of Czech military thinking in relation to motorcycles.

The following weekend he used his regular pass to fetch the 350 Jawa from Brno.

On the Saturday next he drove his superior officer home, returned the Škoda to the depot, then hopped a stone wall topped with barbed wire in the back of the barracks behind which he had his Jawa leaning against a tree. The front gates were always heavily guarded but the perimeter, for some obscure reason, was never secured. Frank rode half the night to get to a circuit in Ústí nad Orlici. Sunday morning he practised; in the afternoon he raced and won. By Monday morning he was back, standing with the rest of the soldiers in the barracks at parade.

Next week he did the same thing.

For about a month everything was quiet. Frank himself kept silent, of course, and so did those of his army buddies who knew what he was doing. The racing officials of SVAZARM or DUKLA simply assumed that Frank had the necessary passes from the military. The lieutenant-general had little interest in anything that wasn't his duty to inspect, and certainly no idea about weekend happenings in the world of sports.

But Captain Konoupek did.

This fine, alert officer of the central Prague barracks had many interests, and listening to sport news on the radio was one of them. He also read the newspapers that reported Frank's results, and he knew that Frank was not in possession of the requisite passes. Captain Konoupek was not amused.

His declaration of war was a short command on the parade ground. It happened in front of about 600 privates one Monday morning after the third race of the season.

"Mrazek! Report to my office."

Frank reported, expecting the worst. It wasn't the worst, but it was K.P. duties for five days, peeling potatoes. The lieutenant-general was assigned another driver during the period, and Frank doubted if he ever noticed the difference.

The next weekend he hopped the wall and went racing again. When he returned Monday morning, Captain Konoupek was waiting on the parade ground, purple in the face.

"Mrazek! To my office! Do you think we don't have a radio around here, you moron?"

Frank knew that the bastard had a radio; he only wondered if it would ever occur to him to post any guards at the back of the compound. But Captain Konoupek contented himself with sending Frank to the brig for ten days on a coloured diet: brown bread, black coffee, dappled soup. He was out of jail, though, in time for the next race.

The battle continued all summer. Most soldiers and non-commissioned officers became aware of it, and everybody took Frank's side. The consensus was that Captain Konoupek couldn't get really tough because it would entail a loss of face or maybe even a charge that he couldn't maintain discipline among his men. Frank, for his part, had a stubborn streak like his father. He was in for two years. He was a kulak's son. Sitting in the brig or sitting on his bunk, what was the difference? What could Captain Konoupek do to him except prevent him from racing? If he obeyed the sons-of-bitches he wouldn't be racing anyway.

Captain Konoupek's purple-faced yell, "Mrazek! Pálka!"—which meant jail as usual—became a standard feature of the Monday morning parade throughout that summer and fall. By winter Frank figured that he had spent almost seventy days in the brig. However, he had also won a few races, and it still hadn't crossed anyone's mind to guard the back walls of the barracks or to inform the race officials that he shouldn't be allowed on the grid. Frank was looking forward to another summer of the same when Captain Konoupek hit upon a diabolical notion.

In happened in late January, about eight weeks before the new racing season was to begin. One morning Frank found himself assigned as chauffeur to the Old Man. The Old Man was the ranking general of the Russian military mission. He was a Second World War veteran, a big-shot war hero, a mad-dog ex-flier.

The Old Man was bad news. No driver, so far, had managed to keep out of a court martial and a sentence of two to three years in the coal mines after a month or two of chauffeuring the Old Man. Captain Konoupek was said to reserve this assignment for people he especially loathed, those that he desired to bury alive. When the inevitable court martial took place, he would testify against them in person. The charge would be sabotage; and the result, of course, a foregone conclusion. It was a royal shaft, and the beauty of it was that the captain did not have to lift a finger himself to make it happen.

What happened was that the drivers would all smash the Russian general's Tatra, always a brand new Tatra 603, with the Old Man sitting right inside the vehicle. They couldn't help it. The more cautious they were the more likely they were to crash within a short time. It was because of a habit the mad old general had, which was to monitor the speedometer from the back seat.

The needle on the speedo's face had to hover between 90 and 100 km/h. If it fell below, even for a few seconds, the general started hollering, "Davay, davay!" (Let's go!) It was a direct order. If you were a Czech military driver and a Russian general hollered at you to get a move on, you put your foot to the floor.

Possibly, as a former flier, the mad old Russian thought of the instrument as an air-speed indicator. Maybe he was worried that if the needle fell below 90 the Tatra would stall and fall out of the sky. Perhaps it was a dim recollection of some grave war-time event, a conditioned reflex triggered by getting into any moving cabin, some dreadful memory of a dogfight. Alternatively, it was a question of manhood, or maybe of national pride: an old Russian soldier showing some young Czech bloods what it was all about.

The theories were endless, but whatever made the general keep yelling "Davay!", the results were predictable. It was possible to drive fast; it was possible to take chances; but it was not possible to always drive at over 90 no matter what. For instance, one couldn't try rounding Wenceslas Square every day at 90 km/h without crashing into something sooner or later.

Holy shit, Frank thought, as he looked at the assignment sheet. This is the end.

Captain Konoupek knew what he was doing, and he probably considered it funny. He didn't say a word, but his eyes were dancing merrily as he handed Frank his orders. "You like racing, eh, moron," his eyes seemed to say. "Well, here's your chance to go racing every day."

The thing that happened next, however, demonstrated the built-in limitation of all human planning. It was a big lesson for Frank and it stayed with him for a lifetime.

Captain Konoupek wanted to screw Frank and had worked out a foolproof plan to do it. It was perfect. It was logical. It couldn't fail. Frank had to hand it to his commanding officer: he knew what he was doing. Frank thought himself that he was screwed. Some of his pals tried to console him, saying, Well, you're bloody good, you'll get away with it, but Frank just shook his head. Being good had nothing to do with it. The very thing that made a driver good was knowing when to drive fast and when to drive slowly. To be forced to drive like a maniac was to be forced to stop being a good driver.

Monday at dawn he checked out the Tatra from the compound, polished it to within an inch of its life, then with a heavy heart drove it through Prague and pulled up in front of the general's villa. The Old Man lived in a classy area called Letná, in some big house that used to belong to a real capitalist. Frank had barely stopped when the general dashed out the front door, threw himself in the back seat, and yelled, "Davay!" He gave Frank no time to open the door for him, salute, or even to glance at his face.

Well, he was a madman, everybody knew that. What the hell, there was nothing to do except play out the string. Frank put his foot to the floor.

There wasn't much traffic and the Tatra was flying, but even so the Old Man came up to scratch. He must have yelled "Davay!" at least five times between the Letná and the Centrum. He yelled every time Frank backed off the gas for a curve or a pedestrian. He gave the impression of being genuinely, certifiably insane.

Frank could recall a weird experience when he was about fifteen and had won his second or third motorcycle race. People were patting him on the back, but they were also looking at him a little strangely. They were giving him this long, searching look, as if there were something wrong with him.

It was then that it had crossed Frank's mind for the first time that he might be crazy. He remembered standing in front of the bathroom mirror for a while, thinking about it. Was this the face of a crazy person? He tried making little grimaces, baring his teeth, forcing some spittle into the corner of his mouth like a foaming madman—then laughing and shrugging it off. Well, if he was crazy, he was crazy. Except he didn't think that he was.

This Russian: *he* was crazy. That's all there was to it. A weird shell-shocked old bird, with macho gone to his brain. As he got out of the Tatra he gave Frank a little wink, actually looking at him for the first time, then he gathered up his military greatcoat and swept into the Centrum between two saluting guards. Yes, crazy. And there was no question about it, Frank thought, but that this old man would drive him crazy too.

The next few days were a blur of "Davay, davay!" every morning and every afternoon. It wasn't even a matter of Frank worrying about being court-martialled or flipping the Tatra and killing them both. It was a matter of going nuts. Luckily the Old Man wasn't demanding to be driven anywhere during the day just then; he was busy flying some desk at G.H.Q. Frank could requisition as many gas-coupons as he wanted, check in with dispatch, then have the Tatra and the day pretty much to himself.

On the third or fourth morning, having dropped off the old general at the Centrum, he drove to the garage of DUKLA-Prague with shaking hands. He had no particular plan, but thought that a little bench-racing might calm his nerves. Some guys from the track were sure to be around, a few lucky second lieutenants, working on their machines. Lucky bastards doing what Frank would be doing himself if only his father had been a bum all his life instead of working like a dog and saving his pennies.

"Hey, Frank, what's wrong?" one of the fellows asked as soon as he had walked in, before he even had a chance to open his mouth. Great, Frank thought, it must be written all over my face. By the time he finished telling the guys about the mad Russian, a little circle had gathered around him. They were laughing at first, naturally, but then they fell silent because even though it was funny, it was a serious problem.

"Where is the Tatra?" asked one of them, a fellow Frank had seen around but barely knew. He was a DUKLA second lieutenant, too, racing cars.

"It's parked outside."

"Well, bring it in. Drive it over the pit in the back."

In those days there were no hydraulic hoists, at least not in Czechoslovakia. If you wanted to work underneath a car you drove it over a pit. Frank brought in the Tatra and the second lieutenant went to work. "All right, see if this will help," he said when he was finished.

At four o'clock Frank picked up the Old Man. He put his foot to the floor and drove to the general's villa. There was no sound except the rasp of the Tatra's rear engine all the way between Old Town Square and Letná. Not a single "Davay,

davay". At the villa Frank screeched to a stop, jumped out, and opened the door.

The Old Man got out of the car. He threw a glance at Frank, who was standing at rigid attention, his eyes fixed on the general's left ear. Then the Old Man shook his head, said, "Nu, kharasho" (Good enough) and swept into the villa.

The Tatra's speedometer had indicated 90 to 110 km/h all the way.

From that day Frank had no more trouble with the general. He continued driving like a maniac, except fixing the speedo had done most of the trick. A difference of fifteen kms isn't much and few people would feel it in the back seat, but it's still the difference between the possible and the impossible. A good driver, with a bit of luck, could get away with rounding Wenceslas Square at 75 km/h.

This signalled the start of a whole new period for Frank, and to call it a change for the better would have been putting it mildly. Life in the army, like life in general perhaps, can range from the pitiful to the idyllic depending on one's posting. It also depends on one's tastes, nature, and interests of course, but for Frank the idea of driving a big-shot general from place to place was, if not exactly idyllic, congenial enough. Provided the two of them got along.

And the fact was, after the Tatra's speedo had been fixed, Frank and the Old Man got along just splendidly. They couldn't talk much—the Russian had little Czech and Frank had little of any language except Czech—but the beauty of most Slavic tongues is that a person who speaks one can often grasp the gist of whatever is being said in the other. Another lucky thing was that, for Frank to get along with someone, trading speeches wasn't nearly as important as trading fellowship and, as it turned out, the Old Man was cut of the same cloth.

There they were, the two of them, after only a few weeks: driving and bonding together as if there were no tomorrow. Perhaps, Frank thought, it was because they were both crazy. Perhaps because on some level they both understood life to be about the same thing, though what that thing was neither of them would have troubled to put into words. Still, it was an important thing, maybe the most important. Far more important, at any rate, than the fact that one was an old Russian general and the other a teenage Czech private.

The Old Man probably wasn't even old, strictly speaking, though Frank had thought of him as a very old man at the time. Hell, anybody over twenty-five was old and even over twenty he was getting there. The Old Man must have been forty-five or fifty if he was a day, nothing short of a relic, though pretty vigorous for his age. Frank wouldn't have cared to take him on in an arm-wrestling contest. Even when it came to knocking back tiny shot-glasses of vodka Frank could just barely hold his own.

They had similar tastes. Expecting a car to be sparkling clean, for instance. They both liked neat uniforms: green for the army, blue for the air force, or shiny leather jackets when in a civilian mood. The general insisted that his driver should always be dressed as he was, including muftis, which was just fine with Frank. They had similar tastes in driving, drinking, and smacking their lips over pretty girls. Eating? In Frank's own view, they ate like pigs. What Frank especially liked was that, perhaps for the first time in his life, he seemed to be the more prudent of two. He'd take chances driving, but not lunatic chances. He'd wolf down three big sausages with bread, but shake his head at the offer of a fourth. He'd rarely drink himself into a stupor, and at the racetrack he wouldn't drink at all. The Old Man, to judge by how he lived (and also by some stories he told), would fly into combat half-pissed and think it was all a big joke.

The general became like Frank's father in many ways, but a different kind of father. At home Starý Franta would smell Frank's breath for alcohol or cigarettes and smack him if he thought that Frank had strayed. With the Russian it was Frank's turn to pour the glass of vodka down the sink. At the same time the older man started to treat Frank like his child. He would babysit him: there was no other way to put it.

The big question, the question of racing, which Frank raised before anything else, turned out to be a breeze. The general roared with laughter, then had somebody write a ***bumashka*** for Frank, a little piece of paper with his permission to race on Sundays. It wasn't even a permission, it was an order. It was an ukase, and Captain Konoupek could only glare at it.

Things were turning out badly for Captain Konoupek. Though he tried to be very careful, one day his cup overflowed and he ended up making things even worse for himself. It wasn't Frank's doing, or not only Frank's anyway, because all four drivers assigned to the Russian top brass were enjoying irksome privileges, sapping the poor captain's morale.

What really brought matters to a head was another junior officer, straight from the academy, only a lieutenant but a real martinet in the making, who couldn't resist screaming at the four chauffeurs one night when they came sauntering in, slightly the worse for wear, at around 3 a.m. Well, what could they do? It was a big party for the Russian brass, and it would have been impolite for the Czech drivers to refuse a drop of vodka. The drivers, two good guys from Prague and one from Brno, just like Frank, tried reasoning with the junior-grade shithead, but he just continued screaming, so in the end one of the Prague guys threw his boot at him.

The boot missed, but most armies get testy about privates throwing boots at lieutenants, even veteran privates at sopping wet lieutenants, so Captain Konoupek couldn't be blamed for thinking that, this time, he had them. The only trouble was that he overplayed his hand. Frank could see what was coming, and he actually felt sorry for the captain. He actually wanted to help him by asking if he could phone the Old Man as the four drivers were being loaded into a van bound for the big central military jail.

"Permission denied!"

"Comrade Captain, just to tell him I won't be around to pick him up in the morning . . ."

"Shut up, Mrazek."

By eleven the next morning Captain Konoupek was in big trouble. ("You don't believe, how shit hit fan" was the way Frank put it in English more than thirty years later.) It took until about eleven o'clock for the Russian generals just to find out where their drivers were. When they did find out, the Old Man himself came to the jail. He not only came, but he stayed in the echoing stone chamber until the military police brought him Captain Konoupek. For a full minute the Old Man glared at him, then said in an icy voice: "Nu, kharasho," after which he pushed the hapless officer into the cell, had the guard shut the door and personally turned the key.

It was interesting. Frank hated Russians of course, like nearly everybody in Czechoslovakia, but that old man was something else. Hating Russians was a necessity, it was vital and natural, but it had nothing to do with that madman in his greatcoat dashing out of his house shouting "Davay pashli!" One could have a whale of a time hating Russians all day, then come face to face with that alcoholic maniac and love him. Not just for this or that; not for a privilege or a pass; maybe not even for letting Frank race (though that would go a long way). Love him, not so much for locking Captain Konoupek into a cell as for—Frank wasn't quite sure how to put this—as for locking Captain Konoupek into a cell with all the more relish because the captain wanted only to kiss his Russian ass.

Frank hated people who liked kissing other people's asses, and the good thing was that the Old Man seemed to hate them too. He did expect military courtesies, but if people tried to lick his boots he'd step on them. Frank, who could salute but had no idea how to tug his forelock, Frank, for whom not going straight to the heart of any matter would come under the despised heading of "politics", had the kind of style that suited the general. Well, if it suited him, it was all the better. The general's style also suited Frank.

For the next ten months they were going through army life like—well, not like two lovers, not like father and son, not like personal friends, but not like master and servant either. Though in a sense that's what Frank became, the Old Man's *sluha* or batman. After he threw Captain Konoupek in jail the general didn't stop at that, he took Frank right out of the barracks and installed him in his own villa. (He even wrote some report about how no imperialists can catch a Soviet general nodding or deprive him of his Tatra 603 and strand him in Letná by putting his driver in jail.) But if Frank wasn't a son, friend, or servant to the Old Man, the closest word to describe what he was would be comrade. Not in the Party bigwigs' sense, but as in comrade-at-arms.

Being a general's batman was, as Frank found out to his delight, like not being in the army at all. As the general's comrade-at-arms he hardly saw any arms or even comrades for the next ten months. The general and his wife, a smiling old lady who encouraged Frank to call her "Babushka", kept throwing leather jackets and boxy Russian-made cameras and God-knows-what-else at Frank as gifts. Human nature being what it is, Frank couldn't resist sauntering over to the barracks once in a while, cigarette in lips, hands stuck provocatively in the pockets of his latest civilian leather jacket, flaunting his status before a much subdued and chastised Captain Konoupek, to say nothing of his new pink-cheeked lieutenant. Frank wasn't stupid and he knew that he was courting disaster by not keeping his head low. He knew that the Old Man could be transferred or recalled with the stroke of a pen in Moscow. The minute it happened there'd be hell to pay, but he still couldn't resist it. Tomorrow would have to take care of itself.

But tomorrow wasn't coming yet. What came instead, as the summer haze was settling over the lancet arches and ribbed vaults of Prague, was pretty Tatyana.

Tatyana was the Old Man's daughter. She wanted to be a teacher of "the science of sports", as it was called in the venerable Leningrad institute where she was studying. It may have been just a tad ponderous to refer to gym teachers this way, but it was the Soviet way and Tatyana couldn't help it. She was not in the least ponderous herself.

She had come to Prague during the summer break, and while it would certainly not be true that the Old Man or

Babushka threw their daughter at Frank the way they had thrown some of their other gifts at him, it was a fact that the general made a point of relinquishing both his Tatra and his chauffeur to Tatyana every day after being dropped off at the Centrum.

"Drive her around until 4."

"Where to, Comrade General?"

"What kind of a driver are you? Show her Prague. Davay!"

It could not possibly have escaped the Old Man that his daughter was a young woman or that his chauffeur was a young man (and a rather handsome young man at that, even in his own estimation). As for Tatyana, she was just smashing. A calm, matter-of-fact girl, not too shy, not too bold, with huge eyes, high cheekbones, fine, curly, light brown hair, and a narrow-waisted, small-boned figure carrying a stunning pair of breasts.

Tatyana's breasts alone could have been decisive for some young men, though not for Frank whose tastes in this regard encompassed only what he could hold in the palm of his hand. What might have been decisive for Frank was the fact that Tatyana liked him and made no bones about it. Soldiers or sailors are generally youths deprived of female companionship around the very time when the functioning of their gonads occupies most of their waking thoughts. As a result, boys in uniform rarely look a gift horse in the mouth.

Frank was a little less eager than most of his contemporaries only because, as a racer and a chauffeur, he was in a somewhat better position to indulge his hormonal imperatives. A better position, in fact, than ninety-nine percent of all young men in Czechoslovakia. The general not only permitted him to race on Sundays, but he let him actually drive himself to the race-track, or anywhere else, in the Tatra.

To drive a Tatra 603 (or any passenger car) in the mid-fifties was "open sesame" for a young Czech. It meant having first call on the pick of the crop of East Europe's budding beauties. True, a car was a plus anywhere, but to make anything like the same impact a young man in America would have had to sail into port in an eighty-foot private yacht.

It wasn't just the sense of status that such magnificent conveyances as automobiles lent to their possessors; it was also their great practicality. In Stalin's part of the world there were no motels at all. Checking into a hotel was nearly impossible for young, unmarried couples (or for anyone else, unless on some official mission). There were no co-ed residences or dorms—the very idea would have made the authorities wince—and even ultra-modern parents would rarely permit any serious hanky-panky under the roof of their own (much too cramped or much too closely monitored) living quarters. This, needless to say, did not result in celibacy on the part of the young people of Eastern Europe, only in having to satisfy their impulses under the most uncomfortable circumstances. To say nothing of the most unromantic.

Dark alleys or doorways were less likely to harbour muggers in Brno or Prague than couples in various intimate phases of their courtship, amazingly oblivious to passers-by. After nightfall, and sometimes even before, solid, old-fashioned, virtually bomb-proof telephone booths would be shaking like leaves all over town. In public parks the benches would be occupied by young men and women, while stray dogs would frolic in the bushes behind them with much the same abandon.

Relative to such accommodation, a Tatra 603 was the epitome of both privacy and luxury. It was a setting sufficiently romantic to sweep even the strictest member of the Young Communist League off her feet, especially as she could always tell herself that the privilege of such a magic vehicle could have accrued to no one but the solidest young comrade with the most promising future. As a rule, she wouldn't be mistaken, either; Frank just happened to be an exception.

On the beach at Máchovo Jezero, not too far from Prague, girls would line up for the few lucky guys with access to Škodas or Tatras: guys like Frank, or the even luckier devil, František Štăstný, later to become quite famous on the international racing scene in the Jawa factory's colours, but back then just a fast DUKLA rider unburdened by a capitalist father. In the pits the fellows would merrily rush from vehicle to vehicle to check for love's evidence—always plentiful, since cars in those days doubled as motels on wheels. The long and short of it was that Frank wasn't too desperate. If high-cheekboned Tatyana was both gorgeous and willing, there were still a couple of other things to consider.

First, the Old Man himself. What if Frank had read him wrong? The general had never actually talked to Frank about Tatyana, not in so many words, and what if he expected his driver to show his daughter Prague and nothing else? Misreading the Old Man in this regard could be fatal; the very thought of it sent shivers running down Frank's spine.

Second, whatever the Old Man expected, he wouldn't expect Frank to just *play* with his daughter, to trifle with her emotions. No father, and certainly no alcohol-fuelled general of a father, would ever expect his daughter to be trifled with. But if the Old Man wanted Frank to be serious, provided Tatyana were willing—well, then there were his own parents to consider.

Not to mention Jarka, his girlfriend back in Brno.

In a sense his parents and Jarka were almost the same thing, because both Frank's mother and Starý Franta doted on the girl. Jarka was steady, pretty, and down-home in every possible way: exactly the kind of sterling daughter-in-law most parents envisage for their sons. Frank's parents were delighted when they first met her, and they probably thought of Frank and Jarka as a "couple" long before either Frank or Jarka did. But it was all just talk, anyway, because Frank was only seventeen and going into the army, and who could tell if he would still want her when he came back or if she, just a few months younger, would still be waiting for him?

Frank was on his fourth or fifth day of showing Prague to Tatyana, still wondering whether or not to make a move, cautiously turning his eyes away from the full-breasted future gym teacher's not too shy, not too bold, but all too unmistakable glances, when the telegram came from Brno. His brother Jiří had died.

Frank kept turning the piece of paper over in his hand, uncomprehending. His kid brother Jiří? How could Jiří have died when he wasn't yet fourteen years old?

The Old Man gave him a *bumashka* right away, two days' compassionate leave, and when he arrived home the incomprehensible news turned out to be quite simple, even mundane, though just as final. It was a domestic mishap, a freak accident. Jiří had been alone in the house, about to take a bath, when the pilot light went out in the gas heater. It happened to be a windy day, and whether the draft came back through the chimney or through a windowsill or a crack in some wall, it extinguished the thin little flame in the bathroom geyser. It was something that happened regularly with old-fashioned gas heaters (except back then nobody considered them old-fashioned) and some people could detect it in time and some couldn't. Evidently Jiří had noticed nothing, and the odourless, colourless gas had poisoned him in the bathtub. By the time father and mother came home he was dead. At first they thought that the kid had just fallen asleep in his bath.

Frank arrived in time for the funeral, attended it in a daze, exchanged a few meaningless words with his parents and his sister who seemed equally stunned, then went to the station to take the train back to Prague. There was nothing more anyone could do.

Jarka came with him to the station, and for some reason chose that moment to make some remark about rumours of Frank doing this or that with some girl in Prague. He pretended to be indignant, of course, but marvelled at her intuition.

"Jarka, don't give me a hard time."

"Well, I just wanted to tell you that two can play at this," she said. "Don't think nobody ever asks me out."

"Yeah, well, look, that's it," Frank said. He may not have been a gallant or tactful boy back then, but he was nothing if not straightforward. "Okay. Good-bye."

On the train he thought at first that Jarka was making it really easy, but then it occurred to him that perhaps she was doing it on purpose. Perhaps she wanted to go with whoever was asking her out, but didn't want to be the one who makes the break. It was simpler to accuse him, to make it all his fault. It was just another example of politics, as far as Frank was concerned, but it hurt his feelings all the same. Why do people have to play games?

Well, there it was. His parents would get over Jarka and Jiří and everything. He would get over Jarka himself. Jiří, too, though Jiří was harder. Frank still couldn't believe that the kid was dead. It seemed only yesterday that Jiří had stuck that spade handle through the front wheel of the little CZ, breaking Frank's arm. If Frank could have caught him then, he would as sure as hell have beat the stuffing out of him.

Back in Prague the general asked only if Frank had made it in time for the funeral.

"Yes, thank you, Comrade General."

"Nu, kharasho," the Old Man said. "Tomorrow at 0600 hours we're at the airport. I'm flying an MiG to České-Budějovice and you're going to show me the way."

It was only a joke, of course, because all Frank could see from the back seat of the MiG trainer was swirling banks of cloud. But the Old Man piloted them smartly to České-Budějovice and then to Bratislava on the Hungarian border. "How's that for speed, friend?" he asked Frank when they got back to Prague in the afternoon.

Whatever the general had to do in the southern part of Czechoslovakia, he certainly didn't need to take Frank along for anything except the ride. He didn't need a batman or a driver, but maybe he thought that Frank could do with a comrade-in-arms after he had lost his brother. The Old Man may have been right because Frank and his parents were too absorbed in their own grief to be of much use to each other just then, and a real loss doesn't even begin to hit a person until days or weeks later. In any event, the ride did take Frank's mind off things while it lasted. He liked the way the Old Man was flying, too, and he would have liked it even more if the general hadn't taken a nip from a flat bottle in the leg-pocket of his flying suit every time before taking off. But what the hell, it was the package. You couldn't expect an old warrior to be like a bookkeeper in uniform, like that flopping flat tire, that balding Czech lieutenenant-general Frank had started out with in the army. Frank would sooner take the crazy Russian, pissed or sober. There were the goodies and there was a price. If you wanted the one, you had to pay the other, that's all there was to life.

And whatever the price was to be, the goodies were still being delivered. In a few days, when the general had to go on some mission again, instead of taking Frank along he thrust an envelope at him. There was money in it, hundred-crown bills, more than Frank had ever held in his hand in one batch.

"Take Babushka and the girl in the car. Go for a week, be back here next Monday."

"Where to, Comrade General?"

"Am I the driver, ***durak***? It's August, even the pigeons are sitting in the shade. Take them someplace where it's cool. Tatra, Matra, Fatra, don't you have enough mountains in this country?"

It was hard to believe it. A kulak's kid from Brno, a private in the Czechoslovakian People's Army, going on an all-expenses-paid luxury holiday in the Czech mountains. Driving a Tatra 603 to the finest, most exclusive resorts, strictly reserved for Stakhanovite comrades and the top Party brass. With a huge-eyed, small-boned, narrow-waisted teacher of sport sciences especially imported from Leningrad for his benefit. What did it all say about the world? Frank wondered. Maybe that it didn't matter much what kind of a world it was, as long as you happened to have an address in it at the top.

Maybe it was all a matter of luck. By rights Frank should have been court-martialled by now. By rights he should be coming off a twelve-hour shift in some coal mine, ticking off another day with a nail on a piece of wood over his bunk. Instead, it was all a holiday. It was like some mushy capitalist movie.

Close-up: his eyes meeting Tatyana's calm smile. Cut: his hand reaching out for her hand and placing it on the shifter. A new angle: driving along the serpentines with his fingers covering her fingers. And later, on a straight stretch of road, his lips brushing against her lips, Tatyana's lips, the lips of a Russian general's daughter. Pan to Babushka dozing peacefully in the back seat.

It was strange—strange enough to made Frank shudder. What was there to predict? What was there to be scared about? Who had ended up in the central military jail, Frank or Captain Konoupek? Who was dead at fourteen, Frank or his kid brother? Was it Frank, taking chances, or was it Jiří, trying to take a bath? Who could calculate, who could figure out how it was all going to turn out? Who could second-guess anything?

In which case you might as well let it happen. Go with the flow. If pretty Tatyana wanted to have a memorable holiday in the Czech mountains, who was Frank to say no to it? As for the coal mines, in case he had misread the Old Man, it was certainly smarter to go to the coal mines for a Russian general's daughter than for putting a scratch on a Russian general's car.

The week in the mountains was great. Strictly separate quarters at night, but Babushka wandered off during the day and dozed a bit in the afternoon. Frank, scantily dressed and suitably aroused, would lie on his back and teach Tatyana how to use a standard four-on-the-floor. First up, second down, third over and up, fourth straight down again. Pull on the top for reverse.

It was an eerie, unreal summer for Frank, that summer of 1956. It was like a fairy tale, filled with plenty of beauty and evil. His brother Jiří, who ought to have turned fourteen, was dead. Frank himself, who ought to have been a miserable, exhausted private in the army, was instead a leather-jacketed playboy, racing motorcycles and sauntering around the high

Tatras with a stunning Russian girl on his arm. In the fall Tatyana flew back to Leningrad, but first she kissed Frank on the lips at the airport right in front of her father, who pretended to see nothing. Driving back to G.H.Q. Frank half-expected the general to have him arrested, or perhaps to shoot him and be done with it, but the Old Man kept quiet and didn't even look at him. When he got out of the car, he slapped Frank on the shoulder like some prospective father-in-law, said "Nu, kharasho," then swept into the building between the saluting sentries.

In October the big disturbances started in Hungary. Frank, still paying no attention to politics, understood only that the Old Man, along with the rest of the Russian generals, was in some sort of jam and had to dash between Prague and Bratislava at all hours of the day and night. They even had to switch from the Tatra to a Jeep. One day, after several nights of almost no rest, Frank had his first accident. It certainly wasn't the Old Man's fault; he wasn't yelling "Davay, davay!" at anybody. In fact, the Old Man was sound asleep in the back when it happened. The trouble was that Frank had also fallen asleep at the wheel. They woke up only after the Jeep had flipped over into a ditch.

It was a blustery afternoon in late October. They weren't hurt, and the general said nothing about sending Frank to the coal mines. Instead, as they huddled in the shelter of some trees by the road waiting for help, he asked Frank to go to the Soviet Union with him.

"For good, Comrade General?"

"What else, *durak*? They're transferring me back to Leningrad. We'll live in clover. Even in the winter you can race on the ice."

It was true, because ice racing was big in Russia, with guys going around the turns at seventy-degree lean angles on their spiked wheels. Frank had often wondered what it would be like.

The Old Man still hadn't said a word about Tatyana, but that was what had to be behind it all. There were plenty of batmen and drivers in Leningrad, nobody needed to import any from Brno.

Though it may have been something else, too, because whatever made the Old Man tick it wasn't anything a man was ever likely to reproduce in his daughter. Well, in some daughters maybe, but not in Tatyana. Tatyana was no tomboy. She was a woman: as calmly, happily, deliberately a woman as any that Frank had ever met. Fathers usually needed a son to see a little of themselves in a mirror, which is what parents like to see, whether they know it or not; and what a madman like the general needed was not just any son, but somebody as crazy as himself. Or, if that was impossible, somebody almost as crazy, like Frank.

Frank could sense all this, though he wasn't putting any of it into words, not even to himself. Reducing things to words was not one of his vices, then or later. He could also sense, without even spending any time to think about it, that he wouldn't be going to Leningrad. Not for pretty Tatyana, not for her father, not for Babushka's kindly smile, and even less for boxy Russian cameras and leather jackets. He wasn't cut out to be a "sluha", a batman, not even a sluha promoted to son-in-law. Not any more than Starý Franta was cut out to be the Russians' machinist in the uranium mines for tins of sardines and jars of caviar.

But it was also instinct that made Frank respond in the only way he could without losing the Old Man's goodwill.

"I can't go with you, Comrade General. My parents have no other son, now that my brother's dead."

It was "politics", but at least it was half-true. In any event, it was the right thing to say. Somebody like the Old Man would have had a reply to a lot of things but he had no reply to this, so he said nothing. It was only much later, after they had got back to Prague, that he asked Frank a very shrewd question. It showed that there was nothing the general didn't understand about the army.

"What's going to happen to you, friend, after I'm gone? They're going to eat you alive."

It was nothing less than the truth. Captain Konoupek would be waiting in the wings, rubbing his hands. New regulations had come into force just then, increasing the length of army service for everybody from two years to two and a half—retroactively, of course. Frank would have nine more months to serve after the general was gone, enough time for the army to make him regret that he had ever been born.

"I've been thinking," Frank replied to the Old Man. "I'm a mechanic, I can service planes. There's a military airport in Brno under the command of a Captain Knapp."

The general handed Frank his new orders, directing him to report to Captain Knapp in Brno, just before Frank drove him to the airport for the last time. Babushka had already left for Leningrad. Frank gave the general a note to give to Tatyana, which had taken him about two nights to compose, though it only consisted of about three lines. The Old Man crammed it into his breast pocket without looking at Frank.

On the tarmac, just before mounting the steps leading into the plane, he turned back.

"Well, *durak*, why do you think I stuck with you as a driver?"

"I didn't crash the Tatra, Comrade General."

"Think again," the Old Man said.

Frank was thinking. "I drove faster than the others, I guess," he said finally, but the general shook his head.

"You dawdled like everybody else," he replied. "But you were the only one with enough brains to do something about it. Eh, smartass?"

He looked at Frank for another second, then broke into a guffaw, waved his hand once, whether in dismissal or in lieu of good-bye, and taking the steps two at a time he swept into the plane.

Frank stood at the gate watching the silver aircraft slowly dissolve into the darkening clouds. Was it possible? That shrewd, vodka-sodden maniac. Could the old bugger have known about the speedo all along?

CHAMPIONS IN LOVE

Down by the salley gardens my love and I did meet;
She passed the salley gardens with little snow-white feet.
She bid me take love easy, as the leaves grow on the tree;
But I, being young and foolish, with her would not agree.
W. B. YEATS

A little girl was skating in the backyard.

Lips pursed in concentration, drawing oblong circles in space with her body, skimming on top of the lumpy, rippling ice.

Her skates were the kind that could be fastened onto regular shoes with a key. They had thick blades under a metal platform which gripped the shoes in a sawtooth vise. Skates of this type were not very expensive. Nor were they very secure. They would come off in a jiffy unless one kept tightening them with the key, but the thick blades inspired confidence.

Skating was fun. A lot of kids in Brno found skates in their stockings at Christmas, which may have given Růžena and Bohumil Dočekal the idea to put one under the tree of their little daughter, Jana, in 1946. Both parents thought of skates as a present, but it was Bohumil who had the idea of swishing a few pails of water over the ground in the back garden, turning it into a miniature skating rink.

Winters are crisp in Moravia, and six-year-old Jana was soon skating almost every day. Nobody showed her how, because in the family nobody really knew. The backyard sloped a tiny bit, so it was easy for Jana to make her way to the top, then slide on her skates to the bottom again.

At first it was exciting. After a few days it also became a little boring. There wasn't much to sliding down a tiny slope in a straight line. Jana started looking around for something else to do, and she discovered the garden bench.

The bench happened to be at one end of the backyard. At the other end there was a pear tree. The backyard was small, just spacious enough for a few rows of beans and tomatoes in the summer, so the distance between the bench and the tree was only a few yards. The tree and the bench gave Jana the idea that if she went around them in a circle, first one way and then the other, skating might become less boring.

Soon she started cutting across her own path, making a figure eight by the tree and the bench. This was more like it. Then she reversed her direction and skated backwards—well, she did it by accident at first, but the next time she did it on purpose. It wasn't long before she tried a clumsy pirouette around the tree. Then she squatted down and tried another.

Nobody showed her how. She had never seen anybody do it.

Jana was a serious girl, so she skated the way serious girls do anything: seriously. It would be getting dark outside, and the Dočekals' little daughter would still be skating around the pear tree in their garden. Neighbours passing by would see her and smile because it somehow put things into perspective. A little girl skating around a pear tree. A peaceful, reassuring sight anywhere, but especially in Brno during the first, tough, post-war winter of 1946.

Next winter Jana started skating in the school yard. At first her parents would take her, but then she'd just walk over with her friends. A lot of kids were hanging around the natural ice rink adjacent to the school, which was used as a tennis court in the summer. No lessons for most of the children, only fun. Jana didn't even know that skating lessons existed, and her parents had other things on their mind. Bohumil, a post office clerk, was about to heed the country's call for more industrial workers. A campaign had just then started to recruit labourers for the nation's war-damaged factories, offering pen-pushers special incentives.

One day an elderly gentleman grabbed Jana at the skating rink, scaring the daylights out of her. He demanded to talk with her parents. Jana thought that she had done something wrong, though she couldn't imagine what. However, good little Czech children did not ask questions when confronted by adults who wanted something, so Jana obediently took the gentleman home.

The man turned out to be a figure-skating coach. He was retired, but he'd still drop by the skating rink once in a while. According to him, the little girl had a gift. He insisted that Růžena and Bohumil should enroll her in the Brno skating club without delay.

The Dočekals weren't scared by the old man, but they were every bit as surprised as Jana. Skating? Gift? The only sport Bohumil had ever pursued was gliding, just as a hobby, though he did hold an altitude record. Figure skating was popular in Czechoslovakia, but Jana's parents hadn't been following it at all.

Still, next year when the Beneš Stadium first opened in Brno, they decided to let Jana join the skating club. It was rather a pain in the neck. The new stadium was almost downtown, a long way from the suburbs where the Dočekals lived. Their street was actually on one of Brno's road racing circuits, not the big Masarykův Okruh, the "Masec", but the smaller Královopolský Okruh. There was also the question of expenses, though Jana's grandfather made things a little easier by presenting her with a pair of skates—real ones this time, permanently built into skating shoes. True, the shoes were about three sizes too big, but Jana was expected to grow into them.

At the club the children were sorted into groups, according to talent, though in 1948 nobody used the word. People instinctively shied away from such non-egalitarian expressions. Somehow "talent" seemed too undemocratic. Even worse, it seemed too mystical. It implied that science and effort had little to do with it, that it couldn't be planned and engineered by officials, that perhaps it was a gift from some higher source. So the children's groups were called "novice" or "advanced".

Jana was put with the "advanced" kids, although she had never had a lesson in her life. She was on her way to make up for it, even if she didn't realize it yet.

She was the youngest kid in the club's winter show next year, travelling all over Moravia and Slovakia. They performed in front of audiences: real people, as opposed to just parents and relatives in Brno. The strangers kept applauding, and while Jana was anything but a frivolous girl, she liked that. There was something about the rhythmic sound of clapping from across the footlights, the enthusiastic applause of people one hadn't even met, the admiration of an unseen public from beyond a curtain of blinding brightness.

Perhaps it was her schedule that made Jana so mature and responsible by the time she turned nine. Imagine a nine-year-old getting up at 4 a.m. every morning, practising until 7:30 on the ice, going to school until 1 p.m., then doing her homework and going back on the ice for evening practice until 7 or 8. Year in, year out, day after day. But there was nothing unusual about it. During the winter months dozens of children at the club were on the same schedule. Jana was always too modest to credit herself with any unusual discipline. Or any unusual anything. It was a long ladder, and a lot of hard-working kids were climbing it with her.

Still, somebody had to get to the top of the ladder, and among her friends in Brno it was Jana who reached it first. She won the club's competition at the age of nine, and was sent to represent her region in Prague. In Prague she came third, a notable achievement. She was now third in the country at her level of figure skating.

On her return to Brno, the hard work began. It was at this point that the club's coaches streamed the children into two groups again: recreational and competitive. Naturally Jana was put in with the competitive group. It meant that if she wanted it and worked hard, she could be an athlete, a real figure skater. Not a "pro" because, just like "talent", the word "professional" was rarely spoken in those years. The fact existed but the word did not. People would skate or swim or whatever, but after finishing school they'd also get full-time jobs. Their employers would be some state enterprises. How much time these state enterprises would expect their amateur athletes to spend at their full-time jobs was, of course, a different question. Some amateurs would show up at their offices or factories mainly to pick up their paycheques. This was something that everybody knew, though saying it out loud would have been considered bad form.

Jana was about fifteen when she was first sent out of the country by the sport authorities. She was sent straight to China.

China may seem pretty far away from most places in Europe; but if China is far away from a small town in England or France, no word would express how far it is from a small town in Czechoslovakia. The moon might be farther, but not by much. For a fifteen-year-old to be sent to China just to show Chinese children how to skate was nothing short of miraculous.

And this was not the end of it, because shortly after Jana came home she was sent abroad again: this time to the Soviet Union, to Moscow and to Kiev. Back in those days the Russians weren't tops in figure skating yet, not by a long chalk, and they wanted to learn very badly. So the Czech sport authorities would send one boy, one girl and a figure skating pair to demonstrate the art to young Soviet skaters, and the girl they selected was Jana.

Before she even turned sixteen! Before, even, her first competition in the national championships (because that only came a season later). And certainly without any strings being pulled on her behalf. The Dočekals had no strings to pull. Father Bohumil stayed an industrial labourer until his retirement. But Jana skated beautifully, and she was a serious, responsible young lady. An extraordinarily good-looking young lady, too, as it happened, though she probably was not aware of it at the time—or altogether happy about it later.

She only knew that she was a serious person. More precisely, she knew right from wrong and took pains to be right whenever she could. For instance, she knew that being an honours student was right. Staying suitably modest while making sure that she skated at least as well as her competitors was right. Being neat, orderly, and accurate was right. Being friendly was also right, as long as one kept one's distance. Being gushy, on the other hand, was wrong. Talking about certain things was very wrong. Talking about history or politics was beyond the pale.

Jana was especially aware of her responsibilities when, at sixteen, she skated in the national championships for the first time. Right from the first year she had made it to the summit: she was among the fifteen or twenty competitors who got past the regional qualifiers and could enter the two national semi-finals. At sixteen, she finished among the top ten.

In 1957, when she was seventeen, Jana came third in the Czechoslovakian championship. She was sent next to the European championships with the Czech team, where she came sixteenth overall but third in free-style figure skating, actually ahead of Milena Tumova who was then the Czech national champion. In the following two seasons, in 1958 and 1959, Jana won the national semi-finals. In the championship itself, she came second to Jindra Kramperová in both years. At nineteen she was the second best figure skater in her country, and, in free-style, the third best in Europe.

That year Jana was the cover of the New Year's issue of Czechoslovakia's national sports magazine, the *Stadion*. Everybody in the country knew of her. People would refer to her as "Dočekalová", the feminine form of her surname. The little girl who used to skate around a tree in her parents' backyard thirteen years earlier had now become a star.

Frank, as it happened, saw Jana's picture on the cover of the sports magazine while lying in the hospital.

It was only a fractured collarbone, nothing very serious. He had been racing in Brno when the front bearing broke. It didn't just seize; it actually broke. The front wheel locked, naturally, and Frank went over the handlebars. So he was now lying in bed, leafing through magazines.

That was when he saw Jana's picture. He had heard of Dočekalová before, of course. Though he had no particular interest in figure skating, Frank followed all sports and Jana was not only a national treasure but a favourite daughter in Brno. Everybody had heard of her. Frank had probably seen her before in some news photos, but this was a big, prominent cover shot. Jana looked quite glorious in this picture, and Frank thought, Well, here's a girl to die for.

For Frank such feelings did not translate into distant notions of ethereal, romantic worship. He was not the adoring-fan type. Knightly devotion was never his strong suit. When Frank felt that here was a girl to die for, in practice it meant that here was a girl with whom it would be nice to make as direct a personal contact as circumstances might allow. A body contact, preferably.

However, at that point in time circumstances did not look promising for a contact of any kind. Jana at nineteen had already reached that rarified sphere that was the stuff of dreams for all athletes—or the stuff of world trips and national magazine covers at any rate. Frank at twenty-three was nowhere in comparison. He had been doing well, both in the army and now back on civvy street, but he was not remotely in the same league. His very sport, motorcycle racing, was not in the same league as hers. Not because it was not as popular—it was popular enough in those days—but because it was not as highly organized or developed. Motorcycling did not even have a national championship series. It was not an Olympic sport. Czechoslovakia could not hope to win a European or world title in motorcycle racing, the way it could in figure skating or in hockey. So, naturally, it could not engage the same patriotic pride or count on the same support from the authorities or the press.

Not only that, but Frank was not as close to the top in his sport as Jana was in hers; nowhere near as close. He was not likely to meet her at a reception for dignitaries: he'd never be invited. Which left only private circles—except there were none. For a few minutes Frank was actually racking his brain for mutual acquaintances—after all, they were both from Brno, then a city of fewer than 300,000 souls—but he couldn't think of any.

Oh well, it was just an idea. Frank was not the type to rack his brain over any one thing for too long. He continued leafing through the magazine, trying to squirm his bandaged shoulder into a less painful position and turning his mind to matters more promising than dream girls in cover photographs.

The situation was to change drastically only a few hours later. That was when a friend of Frank's, a young man named Vladimír, happened to drop by the hospital for a visit. They were chatting about Frank's spill in the race, when Vladimír remarked, apropos of Jana's picture that was still lying on Frank's bed:

"Do you know that spot where you went down? It's right in front of Dočekalová's house."

Frank couldn't believe his ears. "You *know* Dočekalová?" he asked.

Vladimír, of all people, turned out to know Jana quite well. In fact, at that time he was meeting her almost every day. A couple of years earlier Jana had switched from the Institute of Foreign Languages, where she had been studying to be a Russian interpreter, to a night school for a degree in physical education. She was skating during the day and attending classes in the evenings. Vladimír happened to be her classmate.

Frank was never one to beat around the bush. As always, he went straight to the heart of the matter. His next question was "Hey, how can I get to her?"

Later the newspapers made it sound romantic. The way the sportswriters were to tell the story, Jana saw Frank crash in front of her house during a race. She saw him; rushed to take him some flowers in hospital, and that's how they met. It was a nice story, except it wasn't true. Jana was not even in Brno when Frank crashed on the Královopolský Okruh in front of her house; she was competing somewhere abroad. If she had been in town, chances are she would not have watched the race. Jana was not interested in motorsports.

In fact it was faithful Vladimír who kept insisting to Jana that there was this fine young man who was at the end of his wits and would do anything—anything—to meet her. Day after day Vladimír was describing Frank in superlatives to Jana. Not exactly on Frank's instructions, only as a result of Frank becoming a pain in the neck about the subject. He was on Vladimír's back all the time after he came out of the hospital. "Vladimír, I'm going crazy. You must set up something with her" was Frank's refrain to his friend, and certainly the simplest way of setting up something with Jana was to describe Frank as a mixture of a Greek god and a saint. Just in case, Vladimír also added that he was the national champion of Czechoslovakia in motorcycling, which Frank wasn't then, not by any stretch of the imagination.

Eventually this buildup was to work against Frank when he met Jana a few weeks later. It was a classic blind date: saying hello to a complete stranger and then going to a movie with him, and it gave rise to one of those intriguing questions that come up sometimes in human relationships.

If Vladimír had not described Frank as a champion and a god, Jana might never have gone out with him. However, Frank could hardly help disappointing someone who was expecting to go out with a champion and a god—and so he disappointed Jana.

For one thing, Jana was a precise person. If she was led to believe that she was meeting a champion, it was a champion she expected to meet. Not a future hopeful; not someone who was at that point farther from a national title than she herself, but a real champion. A championship was a serious matter for Jana, especially at that time, having just missed out twice by a hair's breadth to Kramperová. It was still a sore point with her. She wouldn't find Vladimír's exaggeration a cute trick. She'd find it something close to misrepresentation.

Moreover, Jana would also expect a god to look a little more god-like than Frank did. Not that he wasn't quite all right as young men go, just that he was not *that* good-looking. Maybe Jana was a bit spoiled when it came to looks. She was used to handsome boyfriends. She hadn't dated much—owing to her schedule, she couldn't even make the social scene in dancing school, like most other girls in Brno—but the four or five young men she had gone out with before had all been hockey players or show jumpers or tennis stars, athletes of the first rank. They were strong in the looks department, whatever else some of them might have lacked. Compared to them, Frank was nothing to write home about. His clothes seemed casual, too—perhaps even a bit sloppy.

Far from it being love at first sight, Frank didn't even make a good impression on her. In those days Jana used to be attracted initially by looks or fame—fame meaning some sporting achievement, as she was not then particularly conscious of other types of fame. True, these alone would not satisfy her. If a good-looking or famous young athlete turned out to be a bore she'd reject him, but looks and fame were still first requirements. They were first hurdles that Frank could not hope to pass.

Even if Frank had been an actual champion, the nature of his sport would have done nothing for him in Jana's eyes. Certain activities, of which motorcycle racing is one, are very much a matter of taste. People tend to love speed contests or hate them; they either find them utterly fascinating or completely tedious. Girls in particular, as Frank had discovered before, were either vastly impressed or wholly unimpressed by racing drivers. Jana, alas, seemed to belong to the second type. She wouldn't voice her opinion—she generally refrained from voicing her opinion on other people's affairs—but it was evident that crouching on noisy, smelly, rapid-moving objects struck her as a rather Neanderthal thing to do. Since society allowed racing it must have been all right and Jana was prepared to tolerate it, but that was about all.

Not exactly the stuff of romance, as Frank ruefully noted. But since he wanted Jana—*she* had certainly impressed him in every possible way—he was not prepared to leave it at that. He couldn't just let it go because he was not made to let things go, no more than a greyhound or a cheetah. Whether at the racetrack or elsewhere, chasing things that tried to get away from him was a reflexive response for Frank. It was at the core of his being. Not that he mistook Jana for a competing motorcycle; but the chase still elicited the same reaction in him. He twisted the grip and took off after her.

It would be harder to say why Jana allowed herself to be caught by such a man, a man who didn't impress her at all. She could never quite understand it herself.

Since Frank kept after her, she went out with him again in February. As he insisted, she also took the train once to see him race in Ústí nad Orlicí, in Bohemia. She noted that Frank won his race (God knows, with Jana watching he would not have settled for anything less even it killed him) and also that he was a friendly, outgoing sort of fellow, quite funny at times, a good mixer. As for racing itself, well, it was interesting. Frank was no doubt good at it, though Jana wouldn't have cared if she ever saw another motorcycle race again. It was certainly not Frank's prowess as a racer that started making him more attractive to her.

Perhaps one thing was his age. At twenty-three, Frank was two or three years older than anyone Jana had ever gone out with before. She found that mildly intriguing. Maturity wasn't really the word for it because Jana herself was far more mature for her age than Frank, but all those earlier hockey players and show jumpers in her life were only kids. Frank, at least, was a young man.

Another thing was the fact that Frank seemed to know what he wanted. Jana couldn't stand people who didn't know what they wanted from life. No matter how good-looking they were or how famous, they irritated her and made her uneasy. She had no time for people with existential anxieties: they just didn't seem right to her.

There was no danger of that with Frank. He was as keen to avoid existential uncertainties—he would have used the word "politics" for them, too—as Jana was. He certainly knew what he wanted: he wanted to race, and he wanted Jana. Whether either of the things he wanted had a future or not was a different question, and not one that Jana was eager to address in her own mind at the time.

Perhaps that was why their big quarrel came almost as a relief to her. March 20 was Jana's nineteenth birthday and Frank had arranged a party for her at a friend's home. Jana liked parties, but this one did not turn out happily. It began well enough, they were celebrating, but as the evening wore on Frank started to get, well, *pushy*. At least that's how it seemed to Jana. Her other boyfriends were always respectful. They were never demanding. They liked Jana for what she was. They admired her for her personality, they respected her as a representative of Czechoslovakian sport and so on; but here was Frank at her birthday party, actually demanding things. He was treating her as if she were *that* kind of a girl, which she certainly was not. So they broke up, then and there, and that was supposed to be the end of it.

It might have really been the end of it, too, because Frank resembled a greyhound or a cheetah only in one sense. In another sense he was quite different. Frank had a lot of pride. Greyhounds or cheetahs are rarely burdened with that.

Pride wasn't the only factor. If Frank was ready to drop Jana after their big quarrel, another factor was his attention span for things not connected with racing. Frank was still in love, as much in love as he could possibly be, but racing came first with him. You couldn't race if you had too many other things on your mind.

When Frank had met Jana in January 1959, he had been out of the army for only a short while. He'd had to stay in uniform for nine more months after the mad Russian general had returned to Leningrad. But those nine months weren't exactly wasted. In a sense the general continued to babysit Frank even from a distance.

Captain Knapp, the local commander in Brno, wasn't sure what to do after the Old Man had ordered Frank to report to him. Good guy as he was, he decided to do very little. He gave Frank a job as a maintenance mechanic, looking after the army's trucks and airplanes in Brno, but he never even bothered to assign him a place in the barracks. Frank could live at home, wear civilian clothes almost as often as he liked, and go to the movies on the weekends. There was no question of him having to ask anybody for a pass. Once when he bumped into Captain Knapp in the foyer of some theatre his commanding officer just looked at him and said, "Excuse me, don't I know you from somewhere?"

More importantly, the captain allowed Frank to race. He not only allowed it, but took it for granted that Frank would be racing for DUKLA-Brno just as he had been before. It was part of the deal. So Frank kept racing and winning. He won often enough for his name to be pretty well known by the time he got out of the army. He *was* a champion; not a national champion perhaps, but a champion all the same. He had even built up a following. Frank had his own fans, spectators who came specifically to see him race. It would have been hard to say if all this was as important for Frank as his love for Jana, but it did take up as much space in his mind. If "man's love is of man's life a thing apart," as the poet Byron had it, it was certainly true for Frank.

The second half of Byron's observation would have been less true for Jana. For her, love was not "woman's whole existence." At that time Jana was definitely a figure skater first and a woman second—yet Byron may not have been altogether mistaken.

For instance if Frank had been asked, even hypothetically, to choose between love and racing, he would have said racing. He would have said so without hesitation (though, as a born optimist, he might have added that it was a stupid question because such choices were unnecessary). However, if Jana had been asked the same question, she may have hesitated. It wasn't something she had ever discussed with her coach. So Jana may have hesitated, then carefully replied that it depended on how much in love she'd be.

Though Frank's answer would have been just as truthful, Jana's answer would have been more realistic. More realistic for her, in any event. Her answer would help to make some sense of what happened next, which otherwise might make no sense at all.

It was Frank, not Jana, who had initiated the pursuit. It was Frank, not Jana, who went after it with all guns blazing. It was Frank, not Jana, who felt that the other one was a person to die for. Yet after their terminal quarrel in March it was not Frank who picked up the phone again. Frank would probably never have picked up a phone to call Jana.

It was Jana who phoned first. She phoned Frank in May to wish him a happy birthday.

Why it was that Jana phoned Frank is a question that has no answer. It is a puzzle that Jana herself could never solve. It is safe to say that she did not phone him just to wish him a happy birthday. The birthday was only an excuse for a thinly veiled suggestion to renew a friendship, and both of them knew it. Knowing it, they could relax and start dating again.

They dated through May, June, and July. Jana went to see Frank race a number of times, and, as train schedules were not always reliable, she would end up spending a weekend with him sometimes at whichever circuit he was racing. This was a big deal, considering the time and the place. Jana and Frank were no longer casual dates but an *item*, and a fairly high-profile item at that. Jana, of course, was a star. Frank, if not exactly a star, was a worthy, romantic escort. Sportswriters in Czechoslovakia were not as gossipy as their counterparts in the West—the style of the People's Socialist journalism was greyer and much less personal—but "human interest" was not totally absent even from their copy. The papers seemed to encourage Jana's and Frank's romance, and by extension so did the sports authorities. (If they hadn't, it could have created all sorts of problems for athlete-couples back in those days: the sports authorities were the only game in town.)

But the authorities voiced no objections—which left Frank's and Jana's parents.

Jana's parents, to talk about them first, did not object to Frank at all. Perhaps they might have if they had been more ambitious types, the kind of people who look for strings to pull or ways to rise in the world, but they were not.

If anything, it was Jana's meteoric career that may have filled Růžena and Bohumil Dočekal with a vague, unspoken sense of unease. They were very proud of her, yes; but could anything good come from all these trips to Western Europe and the Orient? Who could tell where such glossy cover photos might lead a decent, strictly-raised young girl? In a breathtakingly short time Jana had been catapulted into an unexplored, glittering world filled with possibilities for danger, temptation, or envy. Jana's parents loved their daughter and were a little afraid for her. For them, Frank was actually a step in the right direction, a retreat from these unknown, exalted heights. He was the boy next door. He liked sausages and beer. Friendly, outgoing, down-to-earth Frank from Brno seemed to Jana's parents to be a solid link with something familiar.

The reaction of Frank's parents may have been the other side of the same coin. If Marie and František Mrazek did not approve of Jana, it may have been for a similar reason. She was a bit too much of a good thing, wasn't she? Cover photos! Trips to Cologne and Kiev! That boy might be biting off a little more than he could chew.

Frank's own view was that his parents' response to Jana was spoiled by the dregs of their great affection for Jarka, Frank's childhood sweetheart from his pre-army days. Wholesome Jarka, any parent's dream of a daughter-in-law. Hard-working, ordinary, unpretentious Jarka, whom one could already see shopping and cooking and washing clothes and producing children even when she was still a child herself. Sweet, relaxing Jarka, who would be no strain on Frank, unlike some globe-trotting, traffic-stopping, photogenic superstar. A man

1946. "A little girl was skating in the backyard." Jana's first skating picture

1957. Frank winning at Bucovise

Early 1950s. Frank and his brother Jiří

1959. Jana's picture on the cover of *Stadion*

1959. Jana and Frank getting married in August

1960. "A national treasure." Jana skating in the Czech mountains

1958. Frank riding the Hinton brothers' Norton to a 4th place finish in the Czechoslovakian Grand Prix at Brno

1960. Frank's father, Starý Franta, is making adjustments to the home-built Jawa-MV Agusta replica.

1961. Bump-start of typical European-style road race. Note the cobblestones and the crowds. Frank is Number 17.

1962. Frank is chasing Gustav Havel (2) at Rosice.

1962. Starý Franta, Frank, Jana, and friends pose after a race at Olomuc.

1962. Winning at Prachovske Skalý on the Jawa 350

1962. With Jana, after winning at Znojmo

1964. Rosice. Ouch, those high curbs hurt. This is why Frank always hated the number 13.

needed a wife, not a national treasure. His parents never actually used these words but Frank could see that this was the gist of their concern.

Starý Franta used no words at all. He stayed aloof, but Mother Mrazek went as far as to take Jana aside for a private chat when the young people first raised the subject of marriage. At least she thought that their talk in the garden would be private, but when Frank sensed what was coming he monitored the conversation through an open window.

Mother chose to employ a time-honoured method. She tried to scare Jana away by describing her own son as a monster.

"Listen, I beg you, stay away from him," Frank's mother explained to Jana. "I'm saying it for your own good. You don't know Frank, but I'm telling you. He's not a steady boy. When it comes to girls, he's . . . you know. He's not going to make you happy."

But Jana would have none of it. She was a stubborn lady herself, and not the type of person to be intimidated or even influenced by frontal assaults. Perhaps she was not invulnerable to manipulation—few people are—but she would have required subtler methods than any at Marie Mrazek's disposal.

"Mrs. Mrazek, I'm getting married to him, not you," she replied to Frank's mother. "I love him, so we'll get married. Whether you like it or not."

They were, in fact, married by August, though it wasn't as simple as this sentence makes it sound. In spite of Jana's brave words she did have misgivings. She felt—and felt so perhaps for the rest of her life—that she had been rushed into this marriage. Not so much by Frank as by her own parents. And by outside circumstances.

She had been offered a new contract that summer. The national club of Slovakia had invited her to skate in Bratislava.

It was all one country, Czecho-Slovakia, but the Slovakian part was different. It was a separate region with its own language, its own customs, and its own rivalries with the rest of the nation. Just then the rivalries were rarely expressed by much more than a little football hooliganism: Slovakian fans roughing up Czech fans, or vice versa. Still, there was quite a bit of head-hunting going on, especially on Slovakia's part, when it came to sporting excellence. The Slovaks tended to offer big perks to top athletes from all over the country to play in Slovakia's colours.

For figure skaters the best perk was the Bratislava stadium itself. It was a big, covered arena, unlike the open rink in Brno. A skater in Bratislava could practise many more hours every day, and always on excellent ice. The coaching was also first-rate. The Slovaks were prepared to offer Jana a good job in one of their state enterprises. She could work as a draftsperson there, for a decent salary, with as much time off as she needed for practice and competition.

Most importantly, there was the matter of housing.

Housing was no worse in Czechoslovakia than in other socialist countries, which meant that it was sheer hell. In Brno it would have taken Jana and Frank about five years to get an apartment of their own—five years, that is, if they were willing to move into the first one available, regardless of where it was or what it looked like. Meanwhile they would have had to live with one or another set of parents, like nine-tenths of all newly married couples in the country. On the other hand Bratislava held out to Jana the celestial promise of a new, private apartment right away. This, as the wily Slovaks knew, was a bribe few people in their right minds could ever turn down.

Frank was all in favour of the move. Starý Franta was still the best mechanic, but there were other tensions at home. At twenty-three and after thirty months in the army, Frank was ready to be on his own. The Slovaks were happy to offer him a job as well, maybe as a truck driver, which was just fine with Frank. Who cared about jobs, anyway? Life was racing, and he could race in Bratislava as easily as anywhere else.

Jana's father and mother were horrified. Not so much by the move itself as by the notion of Jana moving to another city with a *boyfriend*. Decent girls simply didn't do such things. Not by the Dočekals' standards in any event. Jana's parents had their own circles, and in those circles they couldn't have held up their heads if their daughter had been living, unchaperoned, in a distant city with a boyfriend. Nothing would have made such a thing all right: nothing, except marriage. If Jana and Frank wanted to move to Bratislava together, well, they just had to tie the knot and that was all there was to it.

As a good daughter Jana felt pressured by her parents. To be fair, it wasn't just her parents: Jana herself felt uneasy about living with Frank without being married to him. While Jana did consider her parents old-fashioned, she could sense that if she had a daughter she might repeat to her the same sermon one day that her mother had just delivered to her.

However, there was another kind of pressure on Jana. It was an even harder, more ominous pressure.

By changing clubs for material advantages, for perks, wouldn't she be doing the wrong thing, probably for the first time in her life? Wouldn't she betray the people who had made a skater of her, who had sent her to Russia and China? Was switching clubs for such reasons worthy of a representative of Czechoslovakian sport? And whether it was or wasn't, couldn't some other people, some grasping, envious souls, make the suggestion that it was opportunistic and wrong?

Yes, they could. But if Jana were married to Frank, and if Frank decided to move to Bratislava, then no one could say anything. Then even the most envious souls would have to hold their peace. It would be only natural for Jana to follow her husband. What else was a girl to do? Then she would not be regarded as a sellout or a traitor to her old club in Brno.

Perhaps this was an exaggerated fear. Perhaps Jana needn't have felt pressured into marriage by such a consideration. But in those days people were always looking over their shoulders. A few hands in a few powerful offices held the threads to every person's fate. One enemy whispering a word into the wrong ear was enough for all good things to come to an end or for all bad things to begin.

Jana didn't want to take this risk, and who could blame her? Even Frank, easygoing as he normally was, agreed with her precaution. True, agreeing with her only added up to another argument for what Frank wanted to do anyway, which was to marry Jana, but it was still the wise thing to do. So Frank took up his new job in Bratislava first. He moved there, drove back to Brno only for the weekends, and then in a few weeks they had the wedding.

Though Jana felt rushed, she knew that it was still her own decision. If her marriage turned out badly, she wouldn't blame anyone but herself. Nobody had twisted her arm. She would not have married Frank if she had not loved him by then, though she still couldn't say what it was about him that made her fall in love.

The closest she could come was that with Frank she couldn't always get her own way. He was the first young man not to tiptoe around her. The first man to say anything but: yes, Jana my champion, or yes, Jana, my traffic-stopping beauty. The first man not to jump at her every whim. The others always jumped. It was great, but it soon became boring. Frank seemed to know how to pursue and he knew how to go away, but he didn't know how to jump. When she pushed, he pushed right back, and sometimes it was he who pushed first or harder. It was a new experience for Jana, and it was kind of exciting.

As for Frank's parents, they couldn't believe that Frank would marry Jana until it actually happened. Somehow, for some reason, Mother and Father Mrazek had convinced themselves that the whole thing would just blow over. Frank had naturally invited them to the wedding, which was to be held at the house of Jana's parents, but even with the date set his father and mother still couldn't believe it.

On a Friday night in August Frank drove home from Bratislava for the last time. The ceremony was on Saturday, and there was nothing prepared for it in the house. Nothing, except a last-minute plea for him not to go through with it. When that failed as well, his mother finally gave Frank a gift to take to his wedding. The gift represented her view of the entire affair. It was a salami and a large bucket of potato salad.

7

The Winner's Circle

Looking back on things was not one of Frank's habits, especially when he was a young man. For instance, he would have found it hard to say who taught him to race.

It certainly wasn't his father. Starý Franta was a fantastic mechanic, and he was tops as a business manager—a little too good, in fact, as it turned out later. The old man also knew quite a bit about human nature. He could advise his son on strategy, on tricks of defensive shrewdness, such as never to demonstrate during practice how fast his bike really was. (Once when Frank forgot his father's advice about this it cost him not only a race but his racing machine, as we'll see later.)

The one thing Starý Franta didn't try to tell Frank was how to ride on the track. He'd time Frank's laps carefully, then put away his stopwatch with no comment except "You know best how fast you can go."

In a sense this was true, but in another sense it wasn't. Nobody really knows how fast he can go until he tries, and by that time it may be too late. He'll go too fast and crash or, much more likely, he'll go too slowly and get passed. He'll get smoked by everybody, including the service vehicle picking up the corner workers for lunch. Frank had seen people get confused and drive more slowly on the track than they did on the highway getting there.

It certainly wasn't Frank's ambition to get dusted by service vehicles. Just because there was nobody to teach him didn't mean that there was nobody from whom to learn. Right from the beginning he learned a great deal from old Antonín Vitvar.

Vitvar wasn't really old, except in relation to Frank. In his day he used to beat everybody on his Norton. He beat the official works riders on their factory Jawas. He did that until the sports authorities told him to park his British capitalist machine and switch to racing Jawas himself, whereupon Vitvar just quit. He stayed away from the track until the authorities relented. Then he came back, on his old Norton naturally, and beat the Jawa factory guys again.

Everybody cheered for Vitvar, who wasn't riding Nortons to demonstrate the superiority of capitalist technology, but only because he liked riding Nortons. Just like Frank, Antonín Vitvar despised all politics and only wanted to be left alone to race. Except this wasn't easy at a time when everything, including the style of one's haircut, was interpreted as a political statement. In fairness to Comrade Klement Gottwald (or the two Comrade Antoníns, Zápotoczký and Novotný, who succeeded him as president) there had been some other places where haircuts were mistaken for political statements during the fifties, including America, but at least Americans never ran the risk of being sent to the coal mines for the wrong political haircuts.

In any event, Vitvar became a trainer for SVAZARM. He acted like a father to all young racers, and in the early days Frank tried to see if he could pick up a few tricks from him. Most novices imagine during their early days that it's a matter of picking up tricks—not just in racing but in any field—until they discover that there are no tricks to any art or science. Or rather there are too many tricks. You can't isolate them one by one, or hope to learn any one of them until you've mastered most of the others.

What you can do is resign yourself to learning everything a step at a time. In the case of racing that's maybe a few hundredths of a second at a time. When you add up a whole bunch of things you've learned, they can make you, say, two-tenths of a second faster in some corner.

On a track with five such corners that comes to a second every lap. One second amounts to about sixty to a hundred feet on most tracks. By the end of five laps that can put you more than a football field ahead of where you used to be; and, if where you used to be was more than a football field behind some other guy, it means that now you'll be wheel-to-wheel with him. If you can pick up another two-tenths of a second the next time around, you'll have him dusted.

Until you've learned most of the tricks, the ones you already know won't help you much. You have to train yourself to split your attention between at least half a dozen tasks at the same time (realizing that any one you miss can put you on your ass). You must learn to make decisions fast—say, in half a second or less—and to act on them in the same instant. Since any new decision you make will lead you to some new place on the track, it is likely to call for another new decision a split second later. And, as the human brain can't work that fast, you have to teach your body to make your decisions for you.

Which won't be enough, because the human body can't work that fast either. Teaching your hands and feet to make decisions is necessary but it isn't sufficient. The solution is to become a prophet. You must become a futurologist, with a ninety-nine percent rate of accuracy in your predictions. Your solution is to recognize new things before they happen, because after they happen it is generally too late.

Drivers in ordinary traffic must learn the same thing—they have to learn to anticipate—but racers on the track must do it ten times faster. Time shrinks exponentially as speed increases.

Recognizing things before they happen doesn't call for a sixth sense or for any mystical quality. It calls for concentration. Frank understood this right from the start: unless you kept your mind on what you were doing, unless you *concentrated*, you couldn't race. People who liked daydreaming were better off staying at home. It was not enough to enjoy the idea of being a racer; one also had to enjoy the work that went with it. Some people didn't like the work, so they either crashed and packed it in, or remained content to cruise forever at the back.

Recognizing things before they happen becomes easier with practice. It is simpler to predict things the second time around. But no one can get the benefit of a second time without exposing himself to the risk of a first time, in racing or in anything else, so Frank instinctively put experience where it belonged.

It was a fine line between rushing in madly and being too timid to explore. You had to get fast gradually, but you couldn't take forever. Experience was a bonus that came with seniority, and juniors sometimes had to do without it. Frank could see the value of experience, but he also saw that only backmarkers would sit around and wait for it. Winners had to become experienced the hard way. They had to go for it first, and get experienced later.

The main elements of racing were the gift of prophecy, the ability to concentrate, to make decisions with one's body, but there was also something else. It was a kind of quick-wittedness, a braininess that enabled some people to make the right deductions awfully fast from very meagre clues.

There was a story going around about the great Italian driver Tazio Nuvolari, a champion car and motorcycle racer of an earlier era, who came out of a blind sweeper once at around 120 mph to find his racing line completely blocked by a crash that had happened only seconds earlier. There was no way for Nuvolari to have seen the incident or to have been warned about it by the corner workers. The marshalls hadn't even had time to raise their yellow caution flags.

Everybody expected Nuvolari to pile into the wreckage, but he just tightened his line and went around it. He had no trouble changing his line because he had eased off just before coming into the corner. There had been no reason for him to ease off, but he had done so nevertheless. After the race some journalists asked him why.

Nuvolari wasn't sure himself but he thought about it for a while, then said, "Maybe because I could see that the stands were black."

It made sense. Nuvolari had been leading, so naturally all the spectators in the stands would watch his car come around the corner. Normally the blurred glimpse from his cockpit would show him a patch of white as the spectators' faces turned in his direction. That time he saw a patch of black as the spectators turned away from him to look at the accident. It was enough for Nuvolari to sense that something was unusual and back off. By backing off, he could take a tighter line.

Was it experience? No doubt, but it was also something else. It was a guy who could put two and two together while moving at about one hundred feet a second. He couldn't have taken a full second to figure it out because by that time it would have been too late. He could barely have taken half a second, including lifting his foot off the throttle.

If Frank needed this kind of quick-wittedness to be like Nuvolari or like his childhood hero Frank Juhan, he also needed another quality. He needed that quality even more, because one doesn't become a champion just by knowing when to back off the throttle. One also needs to know when to open it all the way.

One must learn how to corner in top gear without ever shutting off. One must discover, almost, how to defy the laws of physics.

But is this possible?

Physical laws are the same for everyone, champions or backmarkers. They're the same for citizens commuting in their Buicks. The greatest champion in the world can't do a thing about gravity. He can't change the effects of mass, weight, inertia, or centrifugal force. At a given track on a given day there is only so much traction available for a given set of tires. If you load them by a factor that is greater, the tires will start drifting. Increase this force even further (say, by applying more power) or reduce the area of tire contact (say, by a ripple in the pavement or by leaning into the turn at a more acute angle) and the tires will slide out.

The limit, the physical limit, is the same for everyone. It is the same for the fast guys and the slow guys. So are track and weather conditions. Racing machines can be very different in their suspension, engines, steering geometry, and the rest, but that's why racing is split into classes. Within the same class these differences are more modest. In any case, they don't necessarily favour the fast guys.

There are also differences in rider experience and skill. They are significant, but not in proportion to differences in rider performance. Certainly the winner of a race seldom has ten times the skill of a tenth-place finisher. There's only so much operational skill involved in driving, and just about everybody on the track has most of it. Moreover, riders are split into various classes as well. Juniors race against juniors. A rider competing in any given class is about as experienced as the next rider on the grid, give or take a few seasons of racing (which, again, may not be in favour of the faster guy).

There are differences in mechanical "feel". A good development rider, for instance, can unerringly sense a difference of a single pound in tire pressure which most people, including many racers, couldn't. Still, the best development riders, though they're never slow, are not necessarily the greatest champions.

This seems to leave nothing but guts. It is tempting to say that in speed contests the real difference between winners and also-rans boils down to plain courage (and Frank was certainly tempted to believe this). No one can step outside the boundary lines of physics, but the fast guys take it to the limit and the slow guys do not. If you have enough guts to take it to the very edge, you win; and if you chicken out, you lose.

There is only one problem with this theory. On the racetrack almost everybody takes it to (or near) the limit. Most people do anyway, whether they acknowledge it or not. To be precise, most people take it to where they genuinely perceive the limit to be.

There are rarely any chickens on the track. Even the guy who finishes last is going to be a pretty gutsy fellow. If he weren't, he'd sit with the spectators. Once he's on the track he, too, will lean the bike over as far as he thinks he can lean it. He, too, will wait before braking. He will try to convert his energy into heat as late as physically possible. (Which, by the way, may or may not be a smart thing. The brake-until-your-eyeballs-pop-out school of thought is only one of many.) Anyway, he, too, will run his machine as deeply into a corner as he thinks the laws of Messrs. Kepler and Newton permit him. Even the guy who is lapped by everyone will usually keep it wide open until a little red light comes on in his mind signalling the approach of some ultimate, physical barrier.

A world champion can do no more.

All people (except those literally bent on suicide) must stop at the red light that signals physical impossibility in their minds. Every human brain—in fact, every living brain—comes equipped with this warning system. Naturally so, because all life would come to a sudden end without it. People or cattle would go over cliffs or walk into raging torrents without something telling them that it cannot be done. So when that ultimate light flashes, when that final buzzer goes on, everybody and everything must pull back.

Guts or determination can only take somebody as far as that warning, but not beyond. (That's a pretty fair distance, by the way. Most people hear the sound of that warning buzzer maybe once in a lifetime, while racers hear it in their minds five or ten times every lap.) But no amount of courage can take anyone beyond that signal, because it's not just a signal. It is a mechanical stop. It acts directly on people's feet and hands. It's like a governor or a rev-limiter on an engine, independent of the operator's will. No one can twist open the throttle against it because his hand won't obey him. (That's why, incidentally, some modern racing schools advise their students to keep their throttles cracked by slightly turning up the idling speed on their engines—though no one thought of doing that in Frank's days.)

If at that inner signal race leaders will back off as promptly as backmarkers—but even backmarkers won't back off before it—how can race leaders be so much faster? (Faster, that is, than a mere difference in skill or experience might account for.) If it isn't even a matter of guts, then what is it? Can physical laws be suspended for some people but not for others?

Not very likely—though looking at some really fast racers one would almost think so. What accounts for the difference is that there is not one limit, but two. The first exists in the real world; the second in people's minds. Or, as old track wisdom has it: "It's all in your head."

The limit in the real world is the same for everybody. The inner limit is different for each individual. Probably all courageous people take it to the "limit", but only to the limit in their minds. Nobody can take it beyond that.

If the limit in someone's mind is at, or close to, the real physical limit, he may become a champion. If it's not even near it, he will stay behind the pack. Practice may extend his limit by a fair margin, but he is not likely to become really fast. He'll be a yeoman, a mid-pack racer, no matter how skilled, experienced, or gutsy he may be.

This is only speculation, but it is possible that evolution favours human brains with huge safety margins. It is possible, even likely, that a little red light begins to flash in most people's minds long before some real physical barrier requires it. That's why most spectators feel that racers are doing the "impossible" when they sweep around a 110-degree 500-foot radius curve at 120 mph. And that is why the lean angle of a truly great champion seems "impossible" even to many fellow racers.

But it's never impossible. It can't be. If it were impossible, a champion couldn't do it any more than anybody else. No one can do the impossible.

The champion's lean angle is possible, all right, only it registers as "impossible" in most people's minds. It doesn't do so in the champion's mind, so he goes for it. His senses are more in tune with the laws of physics. His safety margin is much narrower. He, too, gets scared, but when he has to, not before—which is why he appears fearless.

"Fearless" human beings may be less successful in evolutionary terms, and perhaps that's why there are only a few of them. People with no margin between their inner limits and the real limits of the physical world may not leave as many offspring over the eons of time as other human types. It stands to reason that a freak whose red light comes on only at the actual edge of a precipice and not a micro-second earlier is more likely to fall. He is more likely to break his neck before he can propagate his genes. Someone who feels that something is impossible long before a physical law makes it really impossible is more likely to live and multiply.

What he is less likely to do is win motorcycle races.

In this sense, motorcycle champions are evolutionary throwbacks. They do what they do not so much because they are more gutsy than ordinary people, but because they do not perceive the risk to be as great.

Viewed in this light, Frank was an evolutionary throwback himself. He'd ease off at his own limit like everybody else—he had no choice—but he soon discovered that his buzzer did not sound until well after most other people's. It wasn't that he was much more daring than his competitors, but that he couldn't see so clearly what the fuss was about. As far as he was concerned, he hadn't been pushing it at all. He wasn't going fast. He wasn't taking any chances.

Hell, he was only cruising.

There were, of course, other fast riders on the track. There were a dozen or more who could give Frank a run for his money. Some of them were foreigners, like his Canadian friend Mike Duff. Or the future world champion Southern Rhodesian, Gary Hocking, on his unbeatable 350 cc MV Agusta.

Duff or Hocking could not only give Frank a run; they'd make him eat a load of dust. Frank couldn't even keep them in sight. Hocking on his MV Agusta existed on some other plane, in a different time warp. When he came to Brno for the Czechoslovakian Grand Prix he simply ran away with the race. There were other competitors to keep Frank on his toes, but racing against somebody like Hocking served as a reminder that, no matter how good Frank was, there were machines and maybe also people out there he might never be able to catch. And even if he caught them tomorrow, somebody new and better might come along the day after. Competition was a never-ending challenge. You always had a chance if you were fast, but never any assurance. You might hang a few laurel wreaths around your neck, but you'd never get to sit on them.

If this was the depressing part, it was also the good part. It meant that there would always be someone with whom to race.

Essentially—except for the start of a race, which is always a mêlée—a rider tends to race with only two people at a time: the one immediately in front of him on the track and the one behind. Sometimes he may pass (or be passed by) a string of other riders, but the real dicing is between riders who are in proximate positions and within less than half a second of each other's lap-times. In a sense, the others don't even count. Within two or three laps most "other" competitors usually either pull far away or are left far behind, and then small groups of three to five people will settle down to race with each other.

There are exceptions, such as a significantly faster rider getting stuck at the start and then picking his way through the entire field, but normally a class of twenty to thirty riders will split into a few groups that challenge nobody except one another. In sprint races of fifteen laps or fewer crews rarely hold up signs, so few if any riders will be aware of their positions until after the chequered flag. Most will neither know nor care where they are. They are intent only on trying to pass the guy who is in front, and to keep ahead of the guy who is behind. All battles are for relative positions. They can be as furious for fifteenth place as for first.

A minority of competitors may fluctuate in performance from meet to meet, show much better form or worse form, but most will stay on roughly the same level throughout a given racing season. Not counting mechanical problems or crashes, those who are among the top ten after the first race are likely to stay among the top ten. Riders who begin the season by bringing up the rear will rarely do much better at the end. Racers might, of course, swap positions *within* these groupings, but seldom *between* them.

This relative lack of mobility among competitors, upward or downward, is another indication that racing is largely a matter of innate talent. Dramatic changes come, if at all, mainly from a change in someone's class or equipment. A middling 500 GP rider may switch to the Formula Two class and win. Someone may go from a very uncompetitive to a highly competitive bike. A mid-pack national rider may turn into a regional hero or vice versa. Otherwise, with few exceptions, seventy percent of what a competitor is all about becomes quite plain after his second or third race, and sometimes even after the first. It is the other thirty percent of his talent that is left for him to hone or polish in slow, painstaking stages, a few tenths of a second at a time, over the next three or four seasons.

Frank was certainly not among the exceptions. He won his second race ever, and from that point on he was almost always among the top five finishers, provided he finished at all. Sometimes he crashed, like many fast people, and sometimes he DNF'd (Did Not Finish) when his bike broke down, but by the time he came out of the army he was pretty much settled in his place. He was one of the race leaders. World-class competitors like Hocking would blow by him, but he'd blow by everybody else, except maybe seven or eight people. They would be his real competition, his actual rivals. It was against those seven or eight people that he would race.

Most of his opponents were factory riders: Franta Šťastný and Gustav Havel for Jawa; or Franta Bartoš, Jirka Kostyr, and Standa Malina for CZ. They and Frank were pretty evenly matched. They could all beat Frank—but there wasn't one among them whom Frank could not occasionally beat. Once or twice he managed to beat Šťastný, the best known of all post-war Czech racers, internationally, and he also beat Havel who, in Frank's view, was even better than Šťastný. Not quite as well known, though, maybe because Havel, like Antonín Vitvar, hated politics. He only wanted to keep his mouth shut and race, so he ended up playing second fiddle to Šťastný.

Once when Havel was faster in practice than Šťastný, Frank noticed that the managers of the Jawa team took his bike and gave it to Šťastný for the race. It was just politics as far as Frank was concerned, but pretty disgusting all the same.

On one occasion the same thing happened to Frank.

Since being a kulak's son stayed with one for a lifetime, Frank remained a kind of second-class citizen in Czechoslovakia's official world of sports. He could never hope to become a works rider for Jawa or CZ, or race as a second lieutenant for DUKLA. However, as a fast guy—by 1959 Frank was a SVAZARM championship winner in the 175 cc class twice and once also in the 350 class—he could get the loan of a works bike from Jawa. Factory racing bikes (with rare exceptions) were not sold outright to privateers, but they might be lent for a season if a privateer could be expected to win on them.

Frank's father took the Jawa works bike and immediately replaced all questionable parts, from pistons to coils, with parts of solid West German manufacture acquired through the Mitters. The Mrazeks had kept up their friendship with the father-and-son team from Stuttgart throughout the years, and the two families were always ready to help each other out. By the late fifties Gerhard Mitter started racing Formula Three cars, open-wheel machines of up to 1,000-cc displacement, and whenever he came to race in Brno he won. It wasn't surprising; by that time young Gerhard was well on his way to a European championship in his class.

Equipped with the Mitters' parts, the factory Jawa was soon flying. Without meaning to, Frank ran top practise times on it before a race in Bohemia.

Practise times were important because, unlike most meets under American rules, in Europe it was practise times that determined grid positions. In the United States and in Canada it is usual to run separate heat races to establish starting positions for the main event. In Europe the race officials just stick a watch on everyone during practice, and whoever is fastest gets to sit on the pole.

Starting in the front rows is an advantage. It's not as vital for motorcycles as for cars, obviously, because motorbikes need less room to pass. But it is important all the same. Passing is tough in the sense of being intrinsically dangerous. Also, every moment a racer loses by being stuck behind a slower driver has to be made up for later. The average number of starters in a race is usually 20 to 30 but some grids can have up to 50 to 60 competitors, lined up in rows. Having to fight one's way past a score or more riders in the first few corners just to get a crack at the race leaders is no picnic.

All the same, Frank was never too anxious about his position on the starting grid. He rather enjoyed the notion of coming up from behind. Partly he liked the challenge, and partly he figured that it made for a more exciting race. Like many athletes, Frank had a certain amount of showmanship in his make-up. Trying to treat spectators to a good time was part of his thinking. But the main reason was that, unlike some other racers, Frank was not overly stuck in his groove.

Naturally, having "a line" was as important for him as it was for anyone else. One needed to have a line to go briskly. However, it always seemed to Frank that settling into a groove like a cog railway train made little sense. There were several ways of getting around a corner, and he could usually see most of them at a glance. Passing never struck him as such a big problem. If somebody was sitting on "your" line you could just take a different line and go around him. Who cared if it wasn't as good a line, theoretically, as long as it got you past somebody else. It struck Frank as faintly ludicrous to see racers lined up in corners behind each other like goslings behind Mother Goose.

Once Frank shouted something at one of his friends as he passed him and a few others in a line-up of this kind.

"What was it you said going into that corner?" his friend asked after the race. "Oh I just said, 'Excuse me, is this the right queue for the movie?' " Frank replied, not without a touch of contempt.

That's why practise times were not so important for Frank. Even lap-records were merely icing on the cake. The only thing that mattered was winning. Frank would much sooner win a race by picking up a position than break the lap-record while chasing the winner on a perfect line. (It happens that people staying on someone's tail turn faster laps than they would if they went for the pass. Breaking lap-records, unlike picking up positions, does call for near-ideal lines.)

Frank's style was also influenced by Starý Franta's view that you should never—never—give away in practise how fast your machine really was. What if it led to some conspiracy or sabotage by some envious soul? Let them find out how fast you are during the race.

This was more than just an opinion on the part of old man Mrazek: it was a phobia. Maybe it came from being traumatized by envious souls sending him to the uranium mines; who could tell? In any event, Frank usually took his father's advice, but on this one occasion he forgot. He didn't *think* he was driving too fast, but he evidently was. Frank broke the lap-record—and at the end of practise two Jawa management types appeared in the pits and took his bike away.

It was the factory's bike—Frank only had it on loan—and they needed it, they said. So they took it, along with all the Mitters' West German parts in it. They gave Frank another Jawa to race instead, a real piece of shit that spat out a valve through the exhaust after a few laps.

It was all Frank's fault, according to Starý Franta. When Frank tried to defend himself ("Why are you yelling at me, I was only cruising!"), his father regarded it as talking back, which of course was a no-no. It was a real no-no, even though by that time Frank had been through the army and was a

SVAZARM champion. He was twenty-two years old, but when he started talking back to Starý Franta in the paddock he got the back of his father's hand as if he were a schoolboy. In front of everybody. In Starý Franta's family no one talked back to Father.

Recalling the episode many years later, Frank could take the philosophical view (coloured perhaps by a touch of nostalgia) that having a stern father "may have been good for something". At the time, however, he wasn't feeling so charitable. He did not relish being slapped in the paddock in front of his friends. It was only one of many things, but it became the start of a gradual change in his relationship with Starý Franta. Not just with Starý Franta his father—*that* relationship had been ambiguous throughout Frank's life—but with Starý Franta his manager and mechanic.

Jana, incidentally, had nothing to do with this. It all happened before Frank had even met her. His parents' doubts about his choice of wife only put the lid on a pot that had been boiling for at least a year before.

For instance, it was always Frank's father who handled the racing partnership's money. He advanced it, and he collected it. In the beginning, of course, there was only money to lay out and very little to collect, but by the end of the 1950s the collectibles became significant.

In theory, motorsports were no more "professional" than other sports. Apart from DUKLA-people (rank: second lieutenant, duties: racing) or salaried works riders (officially paid not for racing but for "testing" factory products) competition resulted in no earnings for anybody. However, racing involved awesome expenses, and in socialist countries there were no private, commercial sponsors. The organizers of racing events recognized this. The organizers were the state's representatives, and the funds they had at their disposal came from the state's budget. If the organizers wished, they could be quite generous with funds allocated for the People's Sport.

It was customary to offer start money, prize money, per diem, and mileage to motorcycle racers. There was also a special purse for the ***absolutní vítez*** or "overall winner" of an event, usually the winner of the premier class race. As a rule this meant the 500 cc class, in which Frank would often come first on his 350 cc Jawa. In Poprad, Slovakia, for instance, this purse amounted to 5,000 crowns as Frank eventually discovered—and this at a time when Frank's salary as a mechanic was about 2,000 crowns a month.

But that wasn't all. Sometimes a number of events were scheduled for the same weekend, so racers could pick and choose. If a racer was popular with spectators the organizers might offer him some money under the table to grace their particular starting grid with his presence. There was no official category on the books for such payments, but the money was there. It was one of many quiet ways in which the free market asserted itself in a socialist economy. It had to be hush-hush, but there was nothing on earth wrong with it. It was just show business: people paying for someone they wanted to see.

Frank was popular with the crowd in Poprad. He was popular on many other circuits. He was popular in Prachovské Skály. The latter was a spectacular track, a kind of Isle of Man TT in miniature, a series of twelve-kilometre laps of twisting roads through the mountains of Bohemia. Just like the famous Tourist Trophy of the British original, the roads had every type of pavement invented by man, from asphalt to cobblestones.

One time when Frank happened to win all his races at Prachovské Skály, he dropped by the shed of the clerk of the course to pick up his envelope containing his start money, prize money, under-the-table money, mileage money, and per diem. He picked it up himself instead of letting his father pick it up on their way home as usual.

He did it pretty much on the spur of the moment, acting mainly out of curiosity. He had no idea how much money would be involved. Frank had never discussed money with his father. In truth, it suited him not to have to worry about anything except racing. It also suited Starý Franta to hold all the purse strings. He liked "sitting on the money", as Frank put it. Frank, who after the army was living at home again, even handed his father his regular monthly paycheque from his job (then fought pitched battles with the old man whenever he needed fifty crowns for spending money).

Now when Frank opened the envelope he found 8,500 crowns in it. More than four months' worth of his salary, in crisp new bills.

Old man Mrazek's fury about Frank picking up the envelope was predictable. What Frank would not have predicted was just *how* furious his father would be. Starý Franta was livid. Worse, he was also stunned.

"How could you do that?" he asked Frank. "Who gave you that idea?"

Frank shrugged—in fact he got the idea from a friend who kept egging him on about all the money he must have won racing—and his father said nothing more to him on the drive home. He was in a silent, black rage. Back in Brno he turned to Frank:

"Will you give me the money?"

Frank came by his stubbornness honestly; he inherited it from his father. They were evenly matched.

"No," he replied. "I'm racing, so whatever I win is mine. You didn't win it. If you need money for parts or anything, you can ask me."

On the merits of this argument Frank was probably half-wrong. It wasn't just "his" money; his father's efforts contributed to it in significant ways. Starý Franta had taken risks—considerable risks, especially for a thrifty man—by investing his capital and time in all the motorcycles, spare parts, labour, know-how and organization of their racing venture. The costs over the years must have been enormous. By any fair accounting at least half of this 8,500 crowns as well as all other prize monies probably "belonged" to old man Mrazek.

However, Frank couldn't see it. He never paid much attention to how things got acquired, organized, or fixed. He concentrated on racing. Like most exalted souls—artists or champions who keep themselves at a comfortable distance from the nitty-gritty of ordinary life the better to devote themselves to their obsessions—he could see only the value of his own contribution. Frank never became a prima donna in his manners because he was too down-to-earth and easygoing for that, but his mental habits were a different story. In his own mind *he* was the show; the rest was only housekeeping.

In any case, some of the money undoubtedly belonged to Frank, and his father had never given him a cent. Starý Franta didn't enjoy parting with money. Getting blood from the proverbial stone would have been child's play in comparison. So the two Mrazeks staked out their positions and prepared to dig themselves in.

The main battle was fought in the courtyard of their home. Some of Frank's friends were present to witness it. The engagement was verbal to begin with, but Starý Franta's battles never stayed verbal for long. He soon escalated it by picking up a twenty-litre army gas can and pitching it at Frank.

By that time Mother Marie was also at the scene, and she reacted as mothers frequently do when they see large army gas cans come sailing through the air at their children: she interposed her own body. Naturally, this aggravated the problem. The gas can would probably have caused little damage to Frank, but as Mother Marie tried to catch it, it broke one of her fingers.

Mother started crying in pain. Starý Franta interpreted the injury to his mate as entirely Frank's fault and totally independent of his own actions. Consequently it filled him with righteous indignation. He abandoned artillery preparations in favour of a direct infantry assault, and grabbed his son to teach him a lesson.

Frank had been taught a number of such lessons in the past, and at that point he felt that he had learned enough. Starý Franta was relying on established custom but, like other bullying fathers, he left out one critical element of his calculation: the passage of time. After twenty-two years it was natural for Frank to grow stronger and for Starý Franta to become feebler, so after a brief tussle it was the older man who ended up on his ass.

At this point Frank's friends scattered, including the young man who had started it all by suggesting that Frank pick up his prize money. It wasn't surprising that they scattered because as Starý Franta rose from the dust, white as a sheet and more enraged than ever, the next logical step would have been murder. Frank himself was fairly sure that his father would now try to kill him.

But Starý Franta did not take the next logical step. Demonstrating that his stubbornness was equalled by a perfect sense of reality, he just shifted the engagement from physical to economic grounds. František Mrazek, Sr., was the owner of Frank's motorcycles—so he picked himself up, dusted himself off, and informed his son that he wouldn't be racing anymore. Later that evening Frank found all his personal belongings neatly piled up in the courtyard. That was that. His father had thrown him out of the house.

This gave rise to several problems, but the most immediate one was the following: the Czechoslovakian Grand Prix was only two weeks away. It was Frank's—and everybody else's—biggest race of the year.

When Frank woke up that morning before his quarrel with his father he had a chance for a good showing in the Grand Prix; even a (remote) chance of victory. By the same evening he didn't even have a motorcycle to ride.

There was no likelihood of old man Mrazek relenting. Frank could understand this because he wasn't a relenting type himself. Such people are probably more commonly found in the caring professions than among Grand Prix racers. Even the option of giving in to the old man wasn't really available at this point. Though grovelling and handing over 8,500 crowns would probably get Frank his motorcycle back for the race, it would almost certainly lose him Starý Franta's respect.

This left only one solution, and his name was Bořivoj Sedlák.

Borivoj was one of the boys in Brno, but unlike most of the others he was a pretty sophisticated fellow. A good ten years older than the rest of Frank's mates, he not only used to race Velocettes but had flown with the Czech squadron for the Royal Air Force during the war. As an ex-RAF flight engineer and a racer, Bořivoj knew British bikes. He spoke the British language. Most importantly, he was friends with two fine Australian competitors, Eric and Harry Hinton, who were about to arrive in town to set up their bikes for the Grand Prix.

The Hinton brothers had raced in Brno before. Everyone had marvelled at their enormous English lorry, a complete travelling workshop, loaded with at least ten interesting British motorcycles in various bits and pieces. Frank, who was holing up with some friends desperately trying to figure out a way to race, was sure that the Hintons could lend him a bike for the Grand Prix if Bořivoj would only speak to them.

Three days and two nights before the race a deal was struck. At Bořivoj's intervention—and in exchange for two hundred crowns and two cases of Pilsen beer—the Australians threw a number of motorcycle parts on the ground in the paddock. They said that the parts amounted to a 350 cc Norton in racing trim.

It may well have been true; Frank had no way of knowing. He had never ridden or worked on a Norton before. He had hardly even seen one close up. The Hintons had no manual. From his Velocette days Bořivoj had some vague idea of British technology (a remarkable though also remarkably idiosyncratic technology by other nations' standards), but even Bořivoj knew nothing about Nortons. The deal was fine—with two hundred crowns and two cases of beer for a racing motorcycle, how can one go wrong?—but the enterprise still looked pretty hopeless.

Frank and Bořivoj gathered up the parts and took them to Bořivoj's garage. There they worked for two days and two nights, with Frank's sister bringing them food-baskets from Mother Marie, without Starý Franta's knowledge, of course. On the morning of the first practice session the Norton emitted some blue smoke, followed by a strange, chugging sound.

It probably was a motorbike after all. Frank loaded it up and took it to the track.

It *was* a motorbike but, even so, Frank had never ridden anything like it before. The fact that it had most controls reversed was the least of it; Frank expected that. But until that day he had raced nothing but two-strokes, mainly twins, and the Norton was a four-stroke single. A motorcycle is a motorcycle, but this thing kept going buff-buff-buff on the track, with Frank being convinced that the motor was about to die any minute. The compression nearly stopped the rear wheel whenever he rolled off the throttle. His practise times were the worst ever. Well, Frank thought, at least nobody was going to take this bike away from him.

He did notice, though, that the Norton was going like hell whenever he stumbled on to doing whatever one was supposed to do to make it go. Perhaps there was a way to drive these things, though he would probably not have enough time to learn how. He barely had time to put some pieces of felt around the back of the engine to soak up the oil the Norton was persistently leaking on to the rear wheel. Then, it was time to go on the grid.

Frank started in last place in the Czechoslovakian Grand Prix. It was the forty-seventh place, to be exact. From where he was he couldn't even see the race leaders when the flag dropped. Still, he had a couple of things going for him, and both were pretty important.

The first was the circuit itself, the Masarykův Okruh. The "Masec", his own home track in Brno. Frank knew every corner and every ripple on this track. He could have ridden around it at racing speeds with his eyes closed.

The second was that, before the race, a couple of his friends caught a glimpse of Starý Franta in the stands. His father had come to watch him. Well, there was simply no way, no goddamn way on earth, that Frank would miss a chance to teach the old man a lesson. There was no way that Starý Franta wasn't going to find out, right then and there, that Frank did not need him to race. That he could earn all by himself every penny of every 8,500 crowns anybody had ever stuffed into an envelope.

And so Frank did. He showed his father. Not by winning the 500 Grand Prix on a 350 motorbike—that, in such an international field, would have been physically impossible—but by taking the Norton to the limit of Newtonian physics. He showed it by smoking several factory riders, and international privateers like Bob Brown and Paddy Driver. He showed it by making Jim Redman on his Matchless G50—Jim Redman!—eat dust. He showed it by starting out in forty-seventh place, and finishing fourth.

It was a stellar performance. Redman, for instance, was five times winner of the Isle of Man TT. There was no way anyone, not even Starý Franta, could fail to acknowledge such a result. He and Frank had a talk on the track right after the race, and the upshot of it was that Frank could move back into the house. He could continue racing his father's motorbikes, pay for the spares he needed, and keep his winnings. From now on whatever Frank won at the track would be his own money.

It was a victory in every sense. People were clapping Frank on the shoulder after the race, shaking his hand, and it was beautiful. Some international riders were winking and nodding at him. Yet when it was time to hand out the prizes everyone's attention naturally turned to the winner's circle.

And Frank was not in the winner's circle. There is no room in the winner's circle except for three guys, and Frank was not one of them.

Considering the circumstances it was a miracle that he was where he was. He could have expected nothing more and no one would have expected anything more from him; but the winner's circle was still the only place that mattered and Frank had to stay outside. The leggy blondes weren't kissing him, the champagne wasn't frothing in his hair, and no one was pushing his bike down Victory Lane. Once the celebrations began and the spoils of victory were being shared out, he became no different from the fellow who finished last. Or from the guys who never even started.

Everything was inside that magic circle. Even a single step outside it, there was nothing. Nothing except a memory of a race. Outside the winner's circle it was just a guy remembering a race, uncapping a bottle of beer, loading up a truck, tying down a bike, and going home.

Leaving the Pack

When a person starts racing he may dream of the winner's circle, but he probably has only one realistic ambition and that is to keep the pack in sight.

By his second season he may expect to ride with the pack. First at club level where it's easier, then among senior, national riders. That is already a fast pack and few people can ride with it.

But those who can are rarely content. They generally want to leave the pack behind.

Leaving the pack is not easy. It is not easy even among juniors. It is downright difficult on national levels of racing—to say nothing of International and Grand Prix racing. It means leaving behind a lot of guys who have the same ambition. It means passing a pack of hungry fellows who do not want to be passed. Including one or two who'd rather die.

After a few seasons of trying to get away from the pack some people stop trying. They don't necessarily hang up their leathers, only shift mental gears. They cut the excess weight of pressure. Many, in fact, really begin to enjoy the track for the first time. They don't get slower; they may even get faster, but they relax and just continue riding their own race.

Riding one's own race is a passion. It is attractive, exhilarating, and honourable.

But every season a handful of riders come onto the grid who have a different kind of passion. People for whom relaxing, in this sense, is not a viable choice. They can relax with a glass of beer perhaps, but when they are on the track they want to win. Unless they do, they are dissatisfied. They take next to no pleasure in the process unless it leads them to the winner's circle.

This handful of people—they exist in every sport, of course—find little satisfaction in having done their personal best. Passing someone gives them no boost unless it's someone they have never passed before. They may be quite realistic; they know that it is impossible to win every race, but once they've come third they want to be second, and once they've come second they want to be first. They feel that until they have won, and then won again at the next highest level, they have done nothing.

Others often call such people "obsessed". The word may be accurate, but the pejorative tone is not quite fair. These athletes only recognize and face a simple, uncomfortable truth. A truth that most people prefer to mask with such bromides as "doing one's personal best". The truth, namely, that not to win is to lose.

Or perhaps that mankind is split into two groups: the obsessed and the losers.

The world looks at it the same way, when you get right down to it. There are no books to record the names of those have done their personal best. The almanacs list nothing but winners. The roll of honour exists for no one but champions. Whether at Wimbledon or at Daytona or at the U.S. Open or at Indianapolis, there are only people who get the title and people who do not. Between the Grand Prix and oblivion there is nothing.

In a few men and women this gives rise to a passion to win. If they're accurately called obsessed, those who lack it must be accurately called losers. No one who lacks this passion ever gets to the top in racing, in any sport, perhaps in any walk of life.

Obviously, having the passion is not enough. One also needs sufficient gifts and sufficient good fortune. Having a passion to win, with gifts and fortune to match, may be a blessing. Having only the passion, however magnificent, is likely to be a curse.

Jana and Frank certainly each had a passion to win. In the fall of 1959 it probably equalled the passion they had for each other.

As they settled down in Bratislava they had reason to believe that they also had the gifts to go with their passions. Whether or not their gifts were sufficient was another matter. They had no way of knowing.

The mere existence of talent is easy to discern; what is harder to tell is how much there is of it. Anybody who comes third in Europe in free-style skating at seventeen like Jana is probably the real thing. So is anyone who comes fourth in the Czechoslovakian Grand Prix at twenty-two like Frank. She or he obviously has talent. But does she or he have enough talent to come first? The pyramid becomes infinitely narrower at the peak. Unlike the question of *some* talent that can be answered at the start of a person's career, the question of *enough* talent can only be answered at the end.

As to the Bitch Goddess of Fortune, she remains an enigma from day to day.

Because of the nature of her sport Jana needed luck twice for each of her events. Frank needed luck only once.

Luck plays a big part in a racer's performance but very little in its evaluation. Whoever gets past the chequered flag first, wins. Perhaps it's not quite so simple, because without luck a racer may never get a ride. He may get no chance to perform in the first place because of some sponsor's or team-manager's personal judgement. In certain instances his performance may be subject to protests. Engines may be torn down to see if they conform to specifications for their class. Tests may be run to see whether or not a racer is using legal fuel. Some racers cheat and get away with it; that's "luck" for them and hard luck for their competitors. There may be occasional judgement calls on the track. One guy may be penalized for some infraction, such as passing on the waved yellow caution flag, while somebody else escapes notice.

On the whole, though, racing is a sport that can be measured by a stopwatch. It leaves only a limited discretion to opinion or politics, and none to personal taste.

Figure skating is the opposite. It leaves everything to opinion, a lot to personal taste, and the rest to politics. Along with diving, dressage, gymnastics or synchronized swimming, it is a sport evaluated on style rather than on time or distance. Style, by definition, is a matter of taste.

Frank could have looked as awkward as he wanted on his motorbike. He could have sat on it backwards, but if he crossed the finish line first he would still have won. Winning for Jana consisted in looking "right" in the eyes of the judges. Though this was not an arbitrary judgement—it would be made according to established technical and aesthetic standards which all contestants could anticipate—it was still a personal decision on the part of some individuals.

Nothing proved this personal element more plainly than the fact that judges rarely agreed with one another. Often they differed not only by a fraction—5.6 for one judge, 5.7 for another—but by a whole lot. A 5.3 performance for a French judge might, in all honesty, look like a 5.8 performance to an American judge. Participating in such a sport would have driven Frank crazy.

Even participating in it by proxy, through Jana, was difficult enough. A person like Frank, for whom any divergence of human opinion would come under the heading of "politics" and accordingly be despised, could not help but fume at a sport that involved a lot of real, honest-to-goodness politics. Politics on two levels: among administrative officials and between nations.

It was suspected that some judges adjusted their results for political considerations. Eastern bloc judges were thought to do so regularly, and even judges from other nations would do it once in a while. Perhaps this was a paranoid suspicion, but if so, it was universal. No one could be certain one way or the other because integrity cannot be tested by a measuring tape.

Jana, though far more philosophical than Frank about the nature of her sport, always believed that it was administrative politics that foiled her first two attempts at the Czechoslovakian championship in 1958 and in 1959.

Working with an excellent coach, Stanislav Suk, Jana had won her semi-finals in both years, finishing each time ahead of Jindra Kramperová. Jana felt that her own performance in the finals was just as good as in the semi-finals. Of course, Kramperová may have surpassed both herself and Jana in the finals, but it was also possible that the sport officials sitting in judgement believed that Jana was too young and not sufficiently well-known internationally. They may have thought that Kramperová would have a better chance to bring home international honours for Czechoslovakia, so they made her the champion. This, at any rate, was Jana's feeling, and although bitterly disappointed, she accepted it.

In fact, it only spurred her on to work harder. Jana was good at sticking at her tasks, in a meticulous, teeth-gritting, compulsive fashion. She had exactly the right personality for a figure skater. Although she preferred free-style skating to compulsory figures (in figures a competitor has to be really precise and accurate), she was herself a perfectionist by nature.

As a little girl, for instance, far from having to be told to tidy up her room she would tidy up even her parents' room (then catch hell because her mother couldn't find her things). When Jana grew up she would not only notice if an ashtray was out of place in her apartment; it would seriously bother her. "Bother" wasn't even the right word: it would pain her. She couldn't stand a mess, and would consider anything a mess that was not arranged in neat, geometric patterns. It was a natural tendency, no doubt reinforced by years of having to cut perfect figures on the ice—and soon after moving to Bratislava it yielded its ultimate result for her under a fine new coach, Hilda Mudra.

In January 1960, within six months of her marriage to Frank, Jana won her first national title. She became the figure skating champion of Czechoslovakia.

When talking to newspaper reporters afterwards Jana found it hard to describe how she felt about her victory. Like a good girl she mouthed all the right platitudes, but inwardly she could only shrug. It was really beyond words. How would *anybody* feel if she became a champion? A national champion in figure skating, a champion in one of the oldest Olympic sports, going back to Britain's Madge Syers in 1908 long before the Winter Olympics even began in 1924. Someone with a real chance to follow in the footsteps of such figure skating legends as Norway's Sonja Henie, Olympic champion three times between 1928 and 1936.

How would anybody feel, moreover, about winning a national championship at exactly the right time? Nineteen-sixty was the year of the Eighth Winter Olympics in Squaw Valley, California. It was tremendous to be at the peak of one's form in the very year when thirty nations were expected to compete in twenty-seven events, the greatest number of events in the history of the Winter Games. To be in Squaw Valley, as Jana was certain to be, and perhaps win a medal for Czechoslovakia. Perhaps, with luck smiling at her, even the Olympic Gold. Suddenly (well, not suddenly because it came after fourteen years of awe-inspiring training) but in a way still *suddenly* it was all within Jana's grasp.

The promised private apartment was not ready when Frank and Jana moved to Bratislava, but neither of them cared very much. As it turned out, it wouldn't be ready until early 1962, almost two years later. An apartment was a big thing, perhaps the biggest thing anyone could have in Czechoslovakia at the time, but since it was such a big deal young couples considered it natural not to move into their own places right away. The two-storey, oval-shaped stadium in Bratislava housed a hotel as well as some offices upstairs, and the hotel provided a few rooms to star athletes who were all waiting for their apartments. Completely free of charge, of course.

Frank thought that some of the other "amateurs" at the hotel smirked when they saw Jana actually leave for her civilian job every day. Nobody really expected athletic stars to do that. But Jana did go to work, partly because as a neat, meticulous person she would have regarded it as untidy to pick up a paycheque without showing up at her desk, and partly because she liked her job. She hadn't done any drawing before, she had never studied it in school, but she really enjoyed it. She found that she had a knack for precise geometric shapes and figures on a sheet of paper as well as on the ice.

Frank also went to work every day, though he had less of a choice in the matter than Jana. He rose with his wife at 4:30 a.m., hung around the skating rink until 7, then went to pick up his truck at the City Parks Department. Jana and Frank usually had their dinner together, often at the stadium's restaurant, and by 8 or 9 p.m. they were in their tiny hotel room upstairs. They considered it a good life.

If the extra perks came through Jana's sport, the extra money came mainly through Frank's. Even with a rent-free hotel room, money would have been tight on their two salaries alone. But what with SVAZARM's start money, per diem, and mileage of 1.50 crowns per kilometre, not to mention the purses for winning or placing, Frank could often pick up 3,000 to 4,000 crowns a weekend at a time when his salary was still only 2,000 crowns a month. He could buy himself a Russian-built Volga car. Of course, spares and repairs for the racing bikes cost a bundle as he soon discovered when he had to look after finances himself, but the profits still equalled or exceeded his salary during the summer and fall months.

At first Frank raced his 150 CZ and his two-stroke 350 Jawa, but soon the Jawa factory sold him—sold him outright—a new four-stroke 350 works racing bike. On that bike Frank, a privateer, became king. Newspaper stories often coupled his name with that of Franta Štǎstný, not just because of their rivalry on the track, but because both racing Franks had married well-known figure skaters, though Jana was much further ahead by then in her standings than Jarmila, the factory rider's wife. (Later Jarmila switched to speed skating and became Czechoslovakian champion.)

God knows, Frank had not married Jana out of any calculation, but being married to a national champion still helped him in his own racing career. It counteracted to some extent at least the fact that he was a kulak's son; that he came from the wrong side of the tracks for the people's Socialist Republic.

Luck had been on Frank's side in other ways as well. Once in a while he'd drop his bike like almost everybody, but on the whole he had no really bad crashes. It was all bruises or simple fractures at most. The one exception came when, in search of the Holy Edge, he decided to have a special racing bike built for himself.

In a way it was all because of Gary Hocking's MV Agusta. Frank could never quite forget the future world champion leaving him in a cloud of dust at Brno. He realized that it was a combination of racer and machine, but while he couldn't do much about the racer (except to try harder), he was resolved to do something about the machine.

Obviously, he couldn't buy an MV Agusta—the enigmatic Italian nobleman's unbeatable racing marque was simply not available in Czechoslovakia—but some SVAZARM guys took pictures of Hocking's bike and said that they could build an exact replica of the frame. Once built, they could fit a 350 DOHC Jawa motor into it. In those days DOHC (Double Overhead Cam) motors were the latest thing so the Jawa engine might match any other. As for Count Agusta's machine, the great thing about it was the frame. If Frank and his pals could duplicate it, they'd have a winner.

The long and short of it was that the DOHC Jawa engine worked just fine but the home-built frame turned out to be a pig. Count Agusta obviously knew something that the SVAZARM fellows did not. Their bike *looked* like an MV Agusta, but on the track it steered like a river barge. It certainly wasn't worth the 25,000 crowns Frank ended up paying for it.

One time, in Trenčín, Frank's slide stuck open in a left-hander. That wasn't the frame's fault, nor was it the frame's fault that it happened to be raining, but Frank always felt afterwards that with a different bike he might have negotiated the turn even in the wet and even with his throttle stuck full open. With the Jawa-MV Agusta replica all he could do was to take twenty-eight metres out of a farmer's wooden fence.

People said later that the farmer was happy. His fence had been in need of repair, and when the race organizers paid him he got to build himself a brand-new fence. Frank was less happy because he had a broken collarbone and a compound fracture in his left upper arm.

The doctors at the Trenčín hospital seemed anxious to operate and they talked about "open reduction" to Frank's father, which meant cutting and then setting the bone, but some local racers were not convinced. "Mr. Mrazek," they whispered to Starý Franta, "these guys here are a bunch of butchers. Don't let them lay a hand on Frank." Frank's father decided to go with the medical opinion of the racers, and loaded Frank into the Volga for the five-hour drive back to Brno. On a trailer attached to the Volga rode what was left of the home-built pig. In spite of the needles and pills the Trenčín doctors had pumped into Frank to fortify him for the trip, it turned out to be the worst five-hour period in his life. There was nothing to be done though. Frank gritted his teeth and thought of the journey as a reasonable instalment on the dues payable for the ecstasy of racing motorcycles.

"Hey, Eduard, we have a good one for you!" Starý Franta shouted, rapping on the wooden shutters of a house, when, after an eternity, they arrived in Brno.

Dr. Eduard Kosinka, roused from his sleep at 3 a.m., looked at Frank's arm under an X-ray machine. Then, as it appeared to Frank, he started pulling at it. He kept tugging it until the bone slid back into place. Frank was climbing the walls in pain, but Dr. Kosinka certainly set the fracture by manipulating the arm and without cutting anything. It was not the first bone he had set for Frank, nor, as it turned out, would it be the last. After a few weeks in a cast, then a few more weeks of walking around with a set of weights that kept rolling away from Frank at the most awkward moments—once he remembered collecting them from under the seat in a crowded tram—the arm healed perfectly. By the end of the season Frank was racing again.

The accident had one unexpected outcome. It resulted in a decision by Frank to split up with his father for good. This time they didn't quarrel; Frank parted with old man Mrazek not so much in anger as in resignation. Starý Franta was a good person, a terrific mechanic, the most painstaking, hardworking human being Frank had ever known, but his terminal avarice was driving Frank crazy. At least, this was the way in which Frank came to interpret what happened.

What happened was this: unbeknownst to Frank, his father had been buying accident insurance on him.

In many countries it would never have been possible to buy life or personal-injury policies on speed contestants (at least, not against such harm as may come to them on the racetrack). In Czechoslovakia it was possible, and it wasn't even very expensive. A few months after his accident Frank received notification that his father, as beneficiary, would be paid 24,000 crowns for the injuries Frank had suffered at Trenčín.

Twenty-four thousand crowns was just about what Frank's wrecked motorbike had cost. His father had paid the premiums on the policy, so he would get all the money. He'd get all the money for the bike as well as for Frank's pain and suffering—perhaps naturally, but very unjustly, as it appeared to Frank.

Starý Franta never offered him a chance to buy into his own insurance, he didn't even mention it to him, and of course Frank would never have thought of insurance himself. Such things simply didn't cross his mind. Perhaps this was the old man's way of getting even with him for 8,500 crowns in the envelope. Whatever it was, it was too much for Frank. He decided to race alone.

There was a good tuner and ex-racer named Vladimír Steiner in Prague. He'd rarely come to the races with Frank, but he'd fix his motors. There were two engines for each bike, and whenever one blew up at the track Mr. Steiner would fix it while Frank was racing the other. It was a straightforward business relationship, strictly fee-for-service. It wasn't anything like a team. After splitting up with his father Frank no longer felt like being in a team with anyone.

Jana did very well at the Olympics in Squaw Valley. She came fourth. It was an eminently respectable finish and no one, perhaps not even Jana herself, expected anything better. From Squaw Valley the Czech team travelled straight to Vancouver, British Columbia, for the world figure skating championships, and Jana did very well there, too. She came fifth. On the basis of her combined results there was no question that in March 1960, Jana Dočekalová-Mrázková was among the top five figure skaters in the world.

And she was still barely twenty years old.

But there's only one big story every year and Jana wasn't it. The big story was U.S. champion Carol Heiss. Heiss, who won the Olympic gold in Squaw Valley as well as her fifth world title in Vancouver. She had already won four consecutive world titles since 1956. She was a headliner, a four times American and five times world champion, and now also a gold medalist in the Winter Games. It was all Carol Heiss, with a record second maybe only to Sonja Henie's in the history of women's figure skating.

For those with a passion to win it's always the top of the pyramid that counts, and in 1960 the top of the pyramid was Carol Heiss. Jana's fourth and fifth place finishes seemed meagre in comparison. Nice, yes; respectable, by all means; close, without a question—close, but no cigar.

Jana would not bring home any medals. No one would play

the national anthem for her. Like Frank at the Grand Prix in Brno, she'd have to stay just outside the winner's circle. Down from those three magic cubes by one single, tantalizing step. It didn't matter if other people saw it differently. Even though everybody back home thought that Jana had done splendidly, *this* was how Jana saw it, and it was the way Jana saw it that counted.

Not that she was depressed. It had been a marvellous trip and she had had a good competition. She had done okay. She was far from being disgusted with herself, only a little disappointed. She concluded that perhaps she hadn't been working hard enough. She'd just have to work a bit harder, that was all. Nineteen-sixty wasn't her only chance. At the age of twenty a competitor can usually count on one or two more chances in most Olympic sports, so she set her sights at 1964. Meanwhile there would be three more European and three more world championships. Jana came home, kissed Frank, distributed some exotic American presents to family and friends, then grimly buckled down to practise.

While Jana was making her big break from the pack, Frank was still running with it. In fact, leaving the pack did not even seem to be in the cards for him.

In a literal sense, of course, Frank had been running away from the pack for years. He had won four SVAZARM championships in Czechoslovakia by the end of 1961. But in terms of international, big-time motorcycle racing, equal to Jana's level of competition in figure skating, Frank was actually a non-starter.

To begin with, like all East European racers, he was in the wrong sport. International and Grand Prix events, then as today, were dominated by the big factories with their works riders and machines. A handful of wealthy sponsors and hobbyists could offer the factories some competition in America and Western Europe, but there were no such creatures in Soviet bloc countries.

Nor could Soviet bloc factories compete with Western automotive technology. Russians, Poles, Hungarians, Roumanians, and Yugoslavs had no cars or motorbikes that could run with Italian, British, or West German road-racing machines. Even French, Belgian, Swedish, Dutch, Spanish, or American equipment was rarely competitive. As for the Japanese, well, they were coming up. Coming up fast, even though until a few years earlier no one had heard of them in racing circles. Still, most competitors in Grand Prix racing tended to ride modified British or Italian bikes, purchased with their own (or their sponsors') money.

East Europeans stayed home and raced against each other. They either ran native-industry machines that would be laughed off the track in international competition, or at best older Gileras, Nortons, BSAs, AJSs, BMWs, or Moto Guzzis. Machines of three, four or more seasons' vintage: racing relics that one could pick up at bargain prices and still race for years. But elsewhere in the world technology was on the move. In world events you couldn't run last year's bike in this year's race, certainly not with any hope of getting away from the pack.

Czechoslovakia and East Germany were near exceptions to this rule. Jawa built some world-class enduro racers, and Jawas could just about hold their own in road racing as well. They (and arguably CZs or East German MZs) were somewhere in the middle of the first rank. Or the top of the second rank; take your pick. In any event, they wouldn't be laughed at. But the Jawa factory was entering only a few world road-racing events until the 1960s, and Frank, the kulak's son, could never hope to be a Jawa factory rider.

It was an insurmountable problem. Except for official factory entries it was unheard of for a racer to be allowed out of Czechoslovakia. As a privateer Frank could not have gone abroad to race even if by some miracle he had managed to wangle a ride from a Western sponsor on a competitive machine.

Frank may not even have qualified to start in Grands Prix outside of Czechoslovakia. The complex rules of FIM, the *Fédération Internationale Motocycliste* that governed international meets, reserved the first 15 entries to the first 15 finishers from the previous season's World Championships. Another 10 riders could enter from the various Continental Championships (three from Europe, two from North America, and another five from the rest of the world), after which FIM gave priority to those riders who had already won points in the current World Championship series. It was only when these lists were exhausted that entries would be accepted from affiliated National Federations from individual countries. If Czechoslovakia nominated anyone to race abroad, it would be a factory rider. Since starting grids were generally limited to 40 to 60 entrants, depending on the track, a rider like Frank would usually be left out in the cold.

Every sport has its pluses and minuses. The biggest minus of Frank's sport was that his performance would depend on factors outside his control, such as nationality, money, and technology. A top Westerner with an MV Agusta could always run away from him. Another obvious minus was that a mistake could hurt or kill him. His own mistake or anybody else's.

Jana certainly had the right nationality for figure skating. She was much less likely to get hurt by her mistakes, and her equipment was fully competitive at world level. On the other hand, unlike Frank, she would only get one kick at the can. A pretty brief kick at that, because she would be judged almost entirely by what she could do in a four-minute routine on the ice.

Frank could blow a corner, and make it up on the next. Jana could afford to blow nothing. She had only one moment a year to be in peak form, and it had to be the right moment. There was no way a skater could make up for an error. She could only hope that her opponent made a worse error in *her* performance.

Frank could become a champion without as much as finishing some races in a series. Or, in theory, without winning a single event. Consistent top finishers can grab championships even when they lose most of their races. Frank might goof in several different ways and still collect enough points over the season to come first in a championship. But Jana could not even goof once in competition. If she did a hundred flawless jumps but missed the one that mattered—the one in front of the judges—it would be game over for her.

This translated into pressure. A lot of pressure, naturally not only on Jana but on her opponents as well, except some people respond to pressure better than others. Jana, as it happened, did not respond to it too well. Pressure bothered her, perhaps because she was such a serious, dutiful, responsible person. Being a perfectionist was her main virtue in a way, but in another way it was the worst chink in her armour.

Even though she looked spectacular and showmanship was among her strongest qualities, she found it hard to act ***and*** compete at the same time. What Jana loved was skating at exhibitions. She felt that the people in the stadium were with her and she was with them. Skating for spectators in ice shows was pure joy. In competition, however, her contact with the audience was a bit mechanical. She was too intense, she was concentrating too hard, and it may have lost her some fractions.

Nevertheless the following year Jana won the Czechoslovakian title again. Then, within a month, she came third overall in the European championships. She took home her first international bronze. It was her best result, and it gave her a tremendous boost. She was all set for the world figure skating competition; but the Bitch Goddess of Fortune had different plans.

On February 15 a Sabena Boeing 707 crashed near Berg, Belgium, killing seventy-three people, including eighteen American figure skaters. The 1961 world championships were cancelled.

It was a tragedy for the American skaters and their families, of course, but also for those competitors who were in top form, ready to take on the world, including Jana. In 1961 she had the confidence. She had just hit her stride, she had finally mastered the pressure, and she proved it to herself by bringing home the bronze medal from Europe. She would have done at least as well, she felt, in the world championships; but now she no longer had the chance. True, there was always next year, but who in 1961 could tell about 1962?

Jana was never a complainer. On the contrary, she almost made it a fetish not to blame anyone but herself. At the same time she was somewhat of a pessimist. She tended to view fortune with suspicion. Unlike Tennessee Williams' heroine, Jana would never rely on the kindness of strangers.

Though she didn't feel that judges or sport officials were out to get her, she was sure that they would never let her get away with anything. The only solution was not to make any errors. Ever. Perfection was the answer.

And so the pressure returned.

It had the effect of making Jana nervous about anything that was not the result of a direct effort on her part. Anything that could somehow be held against her: her perks, her trips, even her good looks. She was capable of enjoying herself, but was almost superstitious about admitting it. What if some resentful people decided to get back at her for having a good time?

She would almost have preferred to be ugly. Not in the sense of being pudgy or chunky because a good figure was part of the aesthetics of figure skating and it did entail some effort, but it might have been better to have an ugly face. Judges would never take away fractions for a pair of "bat ears" or a hideous nose. On the contrary, they might resent someone's effortless, undeserved good looks. Maybe—who could tell?—good looks worked against a competitor. It was for this reason perhaps that being complimented on her looks always made Jana uneasy. (Frank felt differently about the subject. He thought that he was rather handsome and didn't mind at all if people noticed it; but then no judge could alter his lap times one way or the other for his looks.)

In 1962 Jana won the national title again. She was now three times champion of Czechoslovakia. In the European and world championships she finished eighth and seventh.

In 1963 it was almost the same story. It was again Jana Dočekalová-Mrázková, fourth time Czechoslovakian champion, but only fourth and eighth in Europe and the world. By then Frank and Jana were living in their new apartment in Bratislava, a fabulous two-bedroom affair and an object of great envy, but Jana couldn't fully enjoy it. The pressure was too great. Gaining a title proved to be easier than hanging on to it.

She hung on to it, though, in 1964. Jana was now five times Czechoslovakian figure-skating champion. It was not a world record perhaps, but it was a serious achievement. Few people have won national championships five times running. And 1964 was also Jana's target year, the year of the Ninth Winter Games in Innsbruck, Austria. For once Frank was also allowed to go abroad as a private tourist, a spectator, to watch his wife compete. It was a big deal because husbands and wives were rarely permitted to leave the country together. For Frank it was his first trip ever outside the borders of Czechoslovakia.

It was an enjoyable trip. After the games Jana and Frank even spent a few days in West Germany with the Mitters. It was amazing for Frank to see his racing friends in their own habitat, and to see the West, with all its magic technology, sparkling toys, with its fantastic Opels and Mercedeses rampant everywhere. "Holy cow," Frank kept thinking, "this is where it's all happening! This is where they make things go fast! What am I doing back there?"

It was a great trip in every way—except for the Olympics. That was something Jana preferred to wipe out from her memory.

The big skating story in the Olympics that year was Sjoukje Dijkstra, the Dutch girl, who had already won the world title twice and would go on to win it for a third time after taking the Gold Medal at Innsbruck. Jana, alas, was not even a footnote. It was nobody's fault, but the pressure got too much for her and she blew it. The experience was bad enough for her to decide not to compete in the world championship that year at all. What she needed was a few months' rest.

Frank dutifully held his wife's hand, but he tended not to take these things too much to heart whether they involved

Jana's efforts or his own. Pressure never bothered him; he wasn't even sure what the word meant. He was just as dogged in his own way as Jana, but his way was different.

For Frank it was, "Why worry? Why race in your mind? You should only race on the track because nothing happens anywhere else. On the track you chase everybody, but once the race is over it is over. You're in one piece, fine. That's all that matters. There'll be another race tomorrow." Of course, even Frank understood that for Jana there might not be another race.

There was one, though, in January 1965. There was one more competition because Jana made what she later came to regard as the worst mistake in her life.

After the Olympics she received the coveted "Master of Sports" title, the country's top honour for athletes, and was invited to a reception at Hradcany Castle in Prague. The president shook her hand. Well, she could have viewed the award as a retirement honour because it was often given to top athletes just past their peak. She could have picked up the scroll that went with the title and hung it on her wall. She could have framed it and quit a winner, right then and there.

But Jana only accepted the award and chose not to take the hint, if that's what it was.

By then Jana had been feeling on the way down for almost two years. Not as a skater—she knew that she was skating better than ever before—but as a competitor. She couldn't pull it out of the hat anymore when it mattered. The pressure was getting to her. She was doing her best work during practice. After five championships no one would have blamed her for calling it a day, but she was still on top. She was still the best in the country. So, instead of retiring undefeated, Jana elected to go for the title again.

Everybody expected her to win it and she very nearly did.

It was more than a close race, it was virtually a dead heat. Jana skated well, she did a good program, and she ended up with the same marks from the judges as her competitor, a young girl named Hana Mašková from Prague. Jana wasn't even a fraction behind, only the other girl had achieved the same total marks with one extra judge scoring her higher than Jana. It was that close. It was that close, but it meant that Mašková had won the championship. She didn't expect to win it. Everybody was surprised, but no one was as surprised as that young girl.

Jana wasn't surprised, only hurt and angry at herself. She saw the result as the sports officials' way of telling her that it was time to go, and she shouldn't have waited for them to say it. It was the one thing that should never have happened.

Frank was sympathetic and supportive, but in his own mind he was glad that Jana was finally out of it. By then he had come to hate international figure skating. It was all politics, as far as he was concerned. The right kind of people, the ones who wanted to keep their mouths shut and skate, were being pulled on strings by the wrong kind of people—judges, officials, God knows who else—like so many marionettes. "They were just like *loutky*," Frank would say twenty years later, having to search for the word in English, though not for the emotion which still lay close to the surface. "They were just like puppets."

So Jana was out of competition, claimed by the pack that had been yapping at her heels ever since she had left it five years ago. Some athletes try for comebacks, but Jana would not even consider it. In a sense losing the title was a relief for her, and more than that, an opportunity. Finally she could do what she enjoyed doing best, which was skating.

Skating for fun, perhaps for a living, and without any pressure.

A competitor in most Olympic sports usually has only five or six years at the top, but she or he may have another five to ten years to reap a few rewards. An athlete can turn professional in a number of different ways. Since figure skating is a type of dancing on ice, one obvious way for a figure skater to turn pro is to start skating in an ice revue. In Jana's case it was the most desirable way.

She loved ice shows. She had the talent for acting on skates and she had the looks. She may have been nervous in competition but she was never nervous, except perhaps in the very beginning, in front of an audience. Far from being hard work for Jana, pleasing spectators was really a breeze.

Ice shows, of course, are show business. Skating in them requires a peculiar gift, which may or may not coincide with pure athletic talents. Someone who has never placed higher than tenth in competition may be better in an ice show than a world champion. The criteria are different. Gold medalists, just like award-winning movies, are not necessarily the greatest box-office hits. Critics and the public often have different choices.

Two of the world's biggest ice shows, the Vienna Ice Revue and America's Holiday on Ice, felt that Jana had exactly what the public wanted. Within days of Jana's retirement both shows made overtures to her—or rather to the Czechoslovakian Figure Skating Federation, which acted as exclusive agent to all figure skaters in the country.

Unlike private agents, the federation of the state did not feel bound by the interests of its clients, fiduciary or otherwise. It felt bound only by its own interests, so it turned down both offers. The officials did not even bother to convey any of this to Jana. They did not as much as inform her. She found out about the offers she got through some foreign newspaper reporters.

In next-door Austria or in faraway America an athlete might have a few things to say about such a cavalier omission, but Jana said nothing. She didn't even blink. As a good Czech girl she found it natural that she wouldn't be allowed to speak for herself in such matters. Still, since she happened to hear about the offers, she humbly asked an official in the appropriate ministry if it were possible for her to find out what was going on.

The answer she received was that a Czechoslovakian National Ice Revue was being formed just then, and it counted on Jana's services. This, the officials explained, would give her an opportunity to repay the state for all the care and money it had lavished on her. It would be a chance for her to compensate the state for all her training and traveling. Anyway, she wouldn't be expected to do it forever. The kindly state would ask her to do it for only two years. After two years of working in Czechoslovakia's ice show they'd let her accept other offers. (If in two years she still had any offers coming in, the officials might have added, but didn't.)

Jana, of course, had little choice. She could either agree to the proposal or never skate again; but that wasn't the point. The point was that the officials' explanation actually sounded reasonable to her. After all, she *was* trained and treated to trips; she *did* lead a pretty pampered life as a top athlete, so why shouldn't her country expect something in return? If anyone had suggested to Jana that she had never freely contracted to such conditions or that they amounted to usury, blackmail, and indentured servitude, she would have looked blank.

Even Frank wasn't too upset or indignant. It was politics again, but so what else was new? At least Jana would be skating and getting paid for it, which was all that mattered. It was all she wanted to do anyway. Good people never wanted to do anything but skate, race, whatever, and keep their mouths shut.

Given this attitude Jana and Frank might have stayed in Czechoslovakia forever, and they probably would have if it hadn't been for a single incident. Before that incident they had never thought of defecting. They had certainly never planned it. Even after the incident Jana would still not have done it if Frank hadn't twisted her arm.

The new ice revue was being formed in Prague, so it was logical for Frank and Jana to move back to Brno in the early spring of 1965. It was closer, and Jana missed her city, her parents, and her friends. Apart from racing Frank had no career, nothing to tie him to any one place, and as a privateer it was just as easy for him to race from Brno as from Bratislava. The route to most circuits led through Brno anyway.

The plan was for Jana to go to Prague for the rehearsals and the premiere, then join the ice show on the road. The first summer's tour was to take the ensemble to Greece, Yugoslavia, and Italy. Jana signed her contract on the condition that Frank could go with them as driver of the artificial-ice truck. Frank had mixed feelings about it; he did not want Jana to be left alone for a whole summer, but the tour would cut into his racing season. Still, he agreed to do it at least once. It was a chance for an all-expenses-paid trip for both of them.

For some reason Greece was cancelled, which left a six-week run through Yugoslavia and Italy. Jana and Frank were actually in Yugoslavia, in Dubrovnik, on a private holiday when the tour started, so it would have been logical for them to join the ensemble at its first performance in Belgrade. However, the authorities wouldn't hear of it. Instead of making a short hop from the Adriatic coast to the Yugoslav capital, Jana and Frank had to fly all the way north to Czechoslovakia first, then take a three-day train ride from Prague through Budapest back to Belgrade with the whole group. It was socialist togetherness, or rather paranoia. There was one communal passport for the ensemble. The officials had to leave with a hundred bodies from the country at the start in order to take a hundred bodies back into the country at the end. It was bureaucracy running amuck as usual, and while it pissed Frank off it didn't particularly surprise him.

The first part of the tour was relaxed and uneventful. The show was a success in Yugoslavia and Jana was enjoying herself. It was not a big, glamorous production with perfect lights and music like the Vienna Ice Revue, but the skaters were good and everybody was having a lot of fun.

A day before the ensemble was due to leave for Italy, Jana, Frank and three other members of the ice revue's staff were called into the road manager's room in the Yugoslavian town of Opatija. Mr. Roman was a nice guy, father of the four-time world champion ice dancing pair Eva and Pavel Roman, but he had puzzling news. Apparently a directive had arrived from Prague. The show was to go on to Trieste, Italy, but without

Jana, Frank, and three other staffers. The Mrazeks were to stay in Opatija for a week, then meet the group in Ljubljana on its return from Italy. No reason was given for the decision.

Jana and Frank were naturally miffed, but they were also mystified. True, it was standard policy not to let husbands and wives travel together to Western countries—it was "safer", as Mr. Roman delicately put it—but Frank had been with Jana at the Olympics in Innsbruck only the year before. They had both returned, so what was the problem? In any event, when the officials contracted Jana and Frank for the ice show, what did they think would happen? Western tours were part of the package. The ensemble was never even supposed to perform inside Czechoslovakia, except for Prague. It was a road show. It was meant to travel in both communist and non-communist countries.

So Jana and Frank were miffed and mystified, but in the end they just shrugged it off. Who could be bothered figuring out the mysteries of the political mind? If the Czechoslovakian state wanted to treat them for an extra week's holiday in Opatija, so be it. They weren't going to make it a big issue. Maybe if the show had been going across the Adriatic to Venice, Frank would have regretted missing *that*, but Trieste was only a few miles up the same eastern coast. It just happened to be on the Italian side of the border. It was a fine little town, but Jana had been there already and it was nothing Frank particularly wanted to see.

Leaving the two of them behind in Yugoslavia was not *the* incident. The incident that set Frank off, that made him livid with fury and determined to get the hell out no matter what the cost, occurred only after the ensemble returned from Italy. Morover it was a purely psychological incident, not an actual act of any kind. Who can ever tell what sets human beings off?

Frank's mind could endure an injury better than an insult, and he regarded what the ice show did in Trieste as an insult to Jana.

In Italy the Czechoslovakian Ice Revue was advertised by a big poster. Jana and Frank first saw it in Ljubljana when the ensemble brought back a copy for them. Essentially the poster consisted of Jana's photograph with her name splashed across it in huge letters. There was some other stuff written on it, but the dominant part of the poster was Jana Dočekalová.

Well, who else? Jana was the biggest star in the show that summer, the five times Czechoslovakian champion, and she also happened to be the prettiest woman. Who but Jana should advertise the ice revue in Italy? But natural as it may have been, it sent Frank right off the deep end.

So that was the kind of treacherous bastards they were. Jana was good enough to advertise their stinking show, but not good enough to go to Italy. The sons-of-bitches! Pulling the strings, letting the puppets dance for them, as if Jana and Frank were not human beings but lousy numbers. Numbers or little painted dolls they could pull every which way for their amusement. Well, not anymore. They wouldn't pull any god-damn strings any longer, Frank said, because the *loutky* were rebelling. They were still in Yugoslavia, an easy border cross-ing away from Austria, and they were getting the hell out.

Jana cried for two days. She didn't want to go. The poster did upset her, it upset her very much, she agreed with Frank about all that; but defecting was still too drastic an answer. Jana couldn't bring herself to even think about it at first.

She was a different person and looked at many things differ-ently from her husband. To begin with, Jana wasn't crazy about foreign countries. They were fine to visit, but they wer-en't home. Home was the only place where Jana felt comfort-able. Perhaps it was because she had travelled so much, or perhaps because she had always done much better in Czecho-slovakia, but Jana associated her happiest moments with her native land. In foreign countries—well, she did okay in some, but often she felt very humiliated, like during the Innsbruck Olympics. Or at sixteen in her first world championship at Garmisch-Partenkirchen in West Germany when she finished last. Last! That was an experience Jana could never forget, never, not in a million years.

There were also her parents. Defecting meant closing the

door on them, and closing it forever. Back in those days if you left the country without permission you were leaving it for good. They'd never let you in for a visit. If you tried to come back, they'd put you in jail.

If they crossed that border into Austria, Jana cried, she would never see her mother and father again. Frank couldn't expect her never to see her parents because of a stupid poster. Besides, neither of them spoke any foreign language. It wasn't like travelling with a team surrounded by officials and interpreters. What would they do? They might never be able to exchange a word with another living soul.

Frank couldn't have cared less about any of this, certainly not in the mood he was in. His parents—sure, he loved them, but they were history. Languages, well, yes; but words were never that important to Frank. He didn't need too many words to talk to a Mike Duff or to the Hinton brothers or to Gerhard Mitter. He could always communicate with people who raced or fixed motorcycles, and as for other types of people, they never understood him anyway. They thought that he was crazy in any language, including Czech.

Nor did Jana's practical objections cut much ice with Frank. What if it was all unplanned, sudden or hasty? What was so big about planning? Frank's mind was calibrated to speed. On the track if you saw a hole you went through it. At 130 mph if you started worrying about ifs and buts, the race would be over before you ever made up your mind. If chance gives you an opening you take it. That's what racing is about, and why should life be any different?

They had two more shows to do in Ljubljana. While Jana was still crying and saying no, Frank went to a telephone and called a friend, an Austrian racer of Czech background with whom he had raced at the "Masec" in Brno. Later Frank wouldn't tell the guy's name to anyone, not even to Jana. In the official version of their escape he turned his friend into "a couple of Austrians" he happened to meet in a Ljubljana bar. But in fact it was a friend, and Frank talked to him on the open telephone line between Ljubljana and Graz. Careless? Maybe, but Frank figured that none of these Serbo-Croatian turkeys who monitor calls would understand Czech anyway. He asked his friend to drive into Yugoslavia and smuggle them out. Just like that and right away.

The Austrian guy was no slowpoke himself on the track. He was cut of the same cloth as Frank, so he said sure, why not. He'd get into his Opel Kapitän four-seater and be in Ljubljana in a day.

This settled the planning stage, as far as Frank was concerned. It left only the problem of his wife. When they went to dinner after the last show on their last evening Jana was still saying no. She kept sobbing in the little room in the Ljubljana hostel where they were staying with all the others, though she tried to be quiet about it. They had hung towels on the window so no one could see into their room because Frank was already throwing clothes into a suitcase while Jana was still shaking her head.

"You trust these sons-of-bitches?" Frank asked her. "You trust them after this? You trust that they'll let you go skate with the Vienna Ice Revue in two years' time? I'm telling you, they'll do only what pleases them. But if that's how you want to live, fine. If you want to be at the mercy of these bastards, go ahead. Me, I've had it. I'm leaving tonight."

It was then that Jana realized that she had no choice. It was no easy thing for Frank to say that he'd go alone, but once he had said it, he'd do it. And if he did leave, then it would make no difference even if Jana stayed behind. Even if she sacrificed her marriage and stayed in Czechoslovakia the officials would never let her skate in any ice revue that was traveling abroad. People whose spouses defected became pariahs. Her career would be finished. There would only be a tremendous fuss with questions, sermons, and threats, and when it was over she could work as a draftsman in some office for the rest of her life. If she was lucky.

"I hope you know what you're doing," she finally said to Frank. "I hope so, because I just don't know anything anymore."

They sneaked out of the Ljubljana hostel at 1 a.m. when everybody was sound asleep. Frank was carrying a single suitcase with a change of clothes for the two of them. The Opel Kapitän was parked right across the street.

Frank's friend had scouted the border crossing on his way in, and in his opinion it was not heavily guarded. Obviously he couldn't take Jana and Frank in his car, but he would take their suitcase. He would slow just before the border and let the two of them jump out, so that they could immediately disappear in the small ravine next to the road. If they followed the ravine for about three hundred yards it should take them behind the Yugoslavian customs shed straight into no-man's-land between Yugoslavia and Austria. It would be just a field, with the light of the Austrian customs shed visible on the other side. They should make straight for the light and if they bumped into anyone shout "Asylum!" He taught them the German word and Frank and Jana kept repeating it during the drive along the banks of the Sava river to the border.

After passing through Trzic the Opel Kapitän started climbing the impressive twenty-nine percent grade toward the 1,300-metre Loibl Pass. The pass itself was already in Austria. Frank's friend slowed the car to a crawl as they were approaching the border crossing. He had his bright lights on, to make it harder for anyone to see two figures tumble into the ravine.

It was a moonless night, with only a few pinpoints of stars. The ravine led Frank and Jana past the back of the Yugoslav customs building. They were crawling by so close that they could hear people's voices inside. Then they caught a glimpse of red, the taillights of their friend's car as it started moving along the highway, so he had obviously made it through. They could see nothing else though. Nothing of the road itself, and certainly nothing like a light on the Austrian side.

In fact, the ravine started to curve away. It was impossible to be sure, but Frank thought that the damn ravine was curving back into Yugoslavia. He had no idea whether they had reached no-man's-land yet but he felt that the ravine was definitely going the wrong way. They had to get out of it.

He pulled Jana after him and they were standing in some kind of a field. Was this no-man's-land? Was it mined? Were there patrols with dogs, maybe people in watchtowers, guards with machine guns?

Frank could see nothing except the light on top of the Yugoslav border station, still not very distant. At least it gave him a direction to move away from and he started walking across the stubbly ground, pulling Jana by the hand, trying to walk as quietly as possible. It was interesting in a way, because Frank had always thought of himself as a pretty cool customer. He couldn't remember being scared of anyone or anything. Now, as he was picking his way through that field with Jana, he could feel a cold, empty sensation spreading in his belly. He had never felt that before. It was as if a huge hole were opening inside his guts, as if all the inner muscles in his abdomen had vanished, and for a while he thought that he might soil his pants. So that's what it was like, real fear. It certainly wasn't pleasant.

Then, suddenly, they saw a cornfield right in front of them. Tall corn. Almost at the same time they saw a second light. It was pretty far away still, beyond the other end of the field.

They were about midway between Yugoslavia and Austria. But what was in that cornfield? It seemed to Frank a perfect place for a patrol lying in wait with machine guns. That's where he would have put them, anyway.

At this point the pressure became too much for Jana. Her nerves gave way and she started running across the corn. Just running and crashing through it like a tractor. Frank froze, but in another second he couldn't even see Jana anymore, so he took off after her. The two of them must have sounded like a herd of elephants going through that corn, and in the pitch-black night the noise seemed to bounce and reverberate from every direction. It was impossible for people not to hear it. Frank expected flares and searchlights to come up any minute, with dogs barking and soldiers firing warning shots—but suddenly they were at the end of the field, very close to the second light, and there was nothing. Not even the sound of wind. Only total, complete silence, and overhead the same cold pinpoints of stars.

They were probably in Austria. Their friend would be waiting for them in the Opel Kapitän parked just a few hundred yards beyond the Austrian customs shed, but which way was the road? Frank didn't want to go near the light and bump into border guards even on the Austrian side. It was too close for comfort. They spoke no German, and what if the guards decided to escort them back across the border? It was better to find the road on their own.

In fact it was their friend who saw them first as they were still wandering in a scrubby field a little distance from the highway. He had parked the car and come to look for them and found them in less than five minutes, he said. No, he didn't hear them come crashing through the corn. Perhaps they hadn't made as much noise as they had imagined.

They got into the car but Jana could remember nothing about the rest of the journey. She immediately fell asleep and woke up only as they stopped for a cup of coffee, near the outskirts of Vienna. The sun was already climbing in the sky.

9

DEATH IN NÜRBURGRING

Not far from Koblenz, in West Germany, the 4.5-kilometre track of the old Nürburgring circuit resembled a mermaid lying on its stomach. For many years Nürburgring was part of the World Championship circuit for both cars and motorcycles, and, in a different incarnation, it still is. Along with Hockenheim, near Heidelberg, the old mermaid was a frequent site of the *Grosser Preis von Deutschland*, the German Grand Prix. Whether or not the track's layout reminded all racers of some mythical beauty from the sea, few doubted the fatal power of its siren song.

Nürburgring had lured several fine competitors to their deaths, although it was probably no different from most other tracks in this regard. All famous circuits have killed some people at one time or another. Opinions have always varied on the relative safety of different tracks. Racing experts may agree on safety standards in the abstract, but on the question of how some circuits meet or fail to meet them, or even on their particular features, there are many disputes.

For instance post-war track architects and sanctioning bodies were big on chicanes but many motorcycle racers hated them. Chicanes—jogs or kinks built into a straightaway, often just a short distance before a major corner—were designed to reduce racing speeds. The idea was that this would reduce the severity of accidents. This may have been true, except it increased their frequency, according to many competitors. Perhaps chicanes worked for car guys who mainly worried about major, high-speed mishaps, but in two-wheeled competition there were no "minor" incidents. A rider could get hurt in a crash of any kind. It was better to reduce the number of crashes than to have more of them at lower speeds. Chicanes, with some exceptions, seemed like a poor trade-off to many motorcycle racers. (The chicane built at the end of the back straight at Daytona in 1974 was one of the exceptions. Going on to an oval banking at around 170 mph was really becoming too risky.)

Both bike and car drivers agreed that track-side obstacles were killers. Nobody wanted trees, curbs, lamp posts, ditches, concrete pillars, and the rest of the stuff that went with old-fashioned road racing. It was track-side obstacles that were responsible for the gradual replacement of true road racing circuits with man-made tracks. While many fans and racers felt that artificial circuits, built usually in the middle of nowhere, could not recapture the festive excitement of races running through ordinary streets, villages, or country lanes, no one regretted the reduction of needless tragedies.

However, on man-made tracks car drivers often demanded safety features that turned out to be a menace to motorcycle racers. Armco—the straight or curved metal guards for which the great British driver Jackie Stewart campaigned long and successfully—became every bit as dangerous as light standards or trees.

Many people thought that at the 1973 Italian Grand Prix these safety devices were responsible for the first-corner deaths of champions Renzo Pasolini and Jarno Saarinen, the "Flying Finn", one of the greatest motorcycle racers in the history of the sport. On the Autodrome di Monza's Curva Grande, a right-hand sweeper following the front straight and taken in those days at around 140 to 150 mph, Pasolini's 350 cc Aermacchi bounced off the Armco directly into Saarinen's path. Their tragedy (followed by a similar three-fatality accident a short time later in the same corner) led to a complete redesigning of the famous Italian circuit. On the new track Curva Grande was now preceded by Variente Goodyear, a tight left-right-left-right chicane, that had the effect of both slowing down competitors and making them unhappy. Chicanes spoiled the fun and were not the answer, many two-wheel racers felt. What they needed for safety were slip roads and run-offs: flat grassy or sandy areas adjacent to the track.

But Nürburgring in the mid-1960s was still an old-fashioned circuit with old-fashioned hazards, like plenty of trees. Whether or not it was worse than most tracks at the time was a matter of opinion. The great British champion Barry Sheene hated it, but then Sheene also hated the "Masec" and the Isle of Man TT. He rode in the British classic only once, and told fellow racer Alan Cathcart that he had tried to make sure he'd have his World Championships wrapped up before Brno so he wouldn't have to race in Czechoslovakia at all. In that sense, at least, Nürburgring was in good company.

The track is fairly fast, about midway between the fastest and slowest of the Grand Prix circuits, with modern (1980s) winning lap average speeds of just a hair over 100 mph for the Premier 500 cc class. Cars, of course, can go faster. The worst "tree-line" at Nürburgring begins at the back straight, just after many motorcycles shift into third gear following a right-hand corner. This is the fastest section of the course.

Even in the mid-sixties a factory Porsche racing car could reach over 215 mph or 345 km/h on the back straightaway at Nürburgring before braking for the chicane. Many propeller airplanes do not cruise any faster. At such speeds cross-wind becomes a big factor, for planes as well as for cars, but even more for cars for two obvious reasons. One, planes don't try to land at cruising speeds. Two, cars can't be steered in the air.

It wasn't a particularly windy day, though, when they found

the scattered bits and pieces of a big eight-cylinder works Porsche among the trees adjacent to the back straight at Nürburgring. The track had been dry and there were no skid marks on the pavement. The Porsche had been lapping in untimed practice and it simply failed to come around the start/finish line after the seventh or eighth lap. The pit crew didn't even begin to look for it until after the practise session was over. The assumption was that the driver had simply pulled off with some mechanical problem.

There had been other cars practising on the course but none directly in front or behind, so no one had seen the Porsche go off the track. There were no spectators and no corner workers there; they are rarely positioned in the middle of a straightaway. It was an unusual place for a mishap. Nor was there any smoke or dust being raised as the car disintegrated deep among the trees, at least none that the other drivers coming around a few seconds later would have noticed. Of course, at 215 mph drivers tend not to look around much anyway. They keep their eyes peeled for the next braking marker far down the track.

For want of any other explanation they attributed the accident to a sudden gust of cross-wind. It was uncommon but possible for a strong gust to hit a car at an angle that would cause its wheels to lift. The driver, a Formula Two European champion and Porsche factory racer, was an acknowledged young expert but the track wasn't too wide and at 215 mph an airborne, unsteerable vehicle could be crashing through the woods in fractions of a second.

This, at least, was how Frank understood Mr. Mitter's explanation of the accident on the telephone. His friend Gerhard was dead.

Gerhard. A young boy fiddling with his MV Agusta at Brno only a few years earlier. A good guy and a good racer, with a wife, children, plenty of money, looking forward to another season as a works driver for Porsche. Gerhard, now lying somewhere near the confluence of the Mosel and the Rhein, mixed with pieces of high-tech engineering and Westphalian soil—because it was exactly like an airplane crash, everything mashed together, according to his father. One second to the next, whoosh.

Mr. Mitter took it well. That was racing.

10

ICE FOLLIES

Frank was shattered by the news, for a variety of reasons. In addition to the grief he felt over the loss of his friend, it also crossed his mind that he could have been behind the wheel of the same Porsche himself when that gust of wind, if that's what it was, hit at Nürburgring.

Though Frank had never really thought of defecting until he caught sight of Jana's poster that summer day in 1965, the opportunity had in fact presented itself a year and a half earlier. This was when they had visited the Mitters in Stuttgart after the Innsbruck Winter Games.

Gerhard was already a driver for Porsche. He had been doing pretty well in other ways too: he owned a few gas stations with his father and was building prototype cars in his own workshop. He even had a dyno on which to test them. Gerhard flew his own plane, had a beautiful home, and was just as enviably high on the ladder of success in relation to Frank as he had been ten years earlier when he brought his mind-boggling MV Agusta to Brno. But Gerhard was also a good guy. He thought highly of Frank as a racer, and while they were visiting he invited Frank to a test track near Paris to take a works Porsche around for a spin.

It was heaven, wheeling that thing around the track. Even

though Frank had never driven a racing car before he thought that a fellow could get used to it in a hurry. It was not as exciting maybe as motorcycles, but certainly the next best thing. After a few laps he was flying, and when he finally pulled in, Gerhard smiled, had a brief chat with an older man in the pits, then asked Frank if he wanted to stick around as a spare driver for Porsche.

They evidently meant it. They said that he wouldn't be racing for the first season but they'd pay him about 80,000 DM a year—worth maybe $20,000 US in those days—if he stuck around and played second fiddle for the team. His job would be to test cars and to be at hand as a substitute for the premier drivers if the need arose.

Naturally Frank mentioned the offer to Jana as soon as they got back to Stuttgart. He mentioned it half in jest and Jana said no bloody way. She was not half-joking, either. She was deadly serious.

Frank did not insist. Jana was still Czechoslovakian champion then, with an opportunity to stay on top for years, and Frank would never have expected her to throw away a career that had taken her a lifetime to build. Besides, Frank was happy in his marriage. He had no desire to split up with his wife, so he just said thanks but no thanks to Gerhard.

That works Porsche was something else though—and so was the idea of getting $20,000 a year just for playing with a toy like that. In Czechoslovakia, his own country, no factory team had ever made such an offer to Frank.

It was totally different from the situation facing them now, just a year and a half later, as they trundled into a police station in Vienna to ask for political asylum.

In 1964 they had been distinguished visitors. They were greeted with flowers. They had smiling guides and interpreters at their disposal everywhere. Most importantly, they could peek at whatever the wide world had to offer from behind the protection of two valid return tickets. If they didn't like it, they could always go back the way they had come.

Now they had blown up their bridges behind them like a

retreating army. They were refugees, with one suitcase between them, beggars at anybody's mercy. Deaf-mute beggars at that, because they couldn't even explain who they were or what they wanted. Jana felt it much more keenly than Frank, but even Frank wondered how he would ever have made the puzzled sergeant sitting behind his desk in Vienna understand what he and his wife were asking for if his Austrian friend hadn't gone with them to the police station.

But his friend did go with them, and the Austrians wasted no time in granting Jana and Frank asylum. By that evening the news was already on the Munich-based Radio Free Europe that Czechoslovakia's former figure skating champion and her husband had defected. God knows how Free Europe had found out, but, as it turned out later, this was how the ensemble in Ljubljana discovered that Jana and Frank had flown the coop. Apparently poor Mr. Roman even tried arguing with the officials in Prague when they phoned him the next morning: "*What* Radio Free Europe? What are you talking about? How could they have defected when they're right here on the beach somewhere!"

Asylum under such circumstances was almost automatic in those days, but other than letting refugees stay and make the best of it, the Austrians couldn't offer them much else. Not even physical security, because agents from Soviet bloc countries kept nipping in and out of Vienna as if it were Grand Central Station. Defectors, if they were important enough, could half-expect some visitors to try and persuade them to return; persuade them one way or another. As a precaution Jana and Frank gave an "official" address in Vienna to the authorities and to the reporters who interviewed them, then promptly went and stayed someplace else.

They stayed in Austria under these conditions for about six months. This was shortly before Gerhard's accident, and it was during this period that Frank saw his friend for what turned out to be the last time. As a refugee with no travel papers he couldn't, of course, visit the Mitters in Stuttgart at that point, but Gerhard came to race his car in a hillclimbing event near

Salzburg, not far from the German border, and that was where they met.

Gerhard won his race as usual, and the two friends chatted afterwards in the paddock. Maybe Gerhard would or would not have repeated his offer to Frank to drive for the Porsche team, but Frank cut him off before he could raise the subject.

The truth was, he didn't want to be tempted by any offers. He had persuaded Jana to leave her home by the promise that she could skate in a Western ice revue—skate without waiting for years, without depending on the whim of officials—and he was determined to make no plans for himself until Jana's career was settled.

It wasn't just altruism. They were refugees without a pot to piss in. The fees that Jana could command as an ice revue star would be more than anything Frank could realistically expect as a spare driver even from a big factory team. So he told Gerhard that he and Jana were off to the New World, and the rest of their talk was just two friends bench-racing. It was like the old days because being a champion hadn't gone to the Mitter boy's head. He was fun; he could both drive and talk a great race. That evening in Salzburg Frank would have believed anything except that he was seeing Gerhard for the last time.

One thing was certain by then. Wherever Jana and Frank might end up, it wouldn't be in Europe. This was why Frank didn't even want to hear Gerhard talk about the Porsche team. They had to leave Europe because there was no way for Jana to join the Vienna Ice Revue.

It was perfectly logical, the minute they thought about it. The simple fact was that shows like the Vienna Ice Revue travelled regularly to Soviet bloc countries. It was part of their normal touring route. They could only recruit East European skaters with the official blessing of their own countries or else they could never build a show around them. A program built around someone like Jana would risk the arrest of its star in Warsaw, Prague, or Budapest. The same was true of other types of touring shows, from circuses to orchestras. For many defecting East European athletes or artists it meant being caught between a rock and a hard place.

It all made perfect sense as soon as the Vienna Ice Revue people explained it, very regretfully, to Frank and Jana. It made sense, but it left only the big American shows: Ice Capades, Ice Follies, or Holiday on Ice. They were pretty much the only potential employers for Jana in the entire world.

The problem was, she didn't know a living soul in America. She knew altogether two people overseas, both Canadians. They were Otto and Maria Jelinek, Canada's world champion figure skating pair in 1962, who were originally from Czechoslovakia but had emigrated as children after the war. Jana had become friends with them at an international competition and she rather thought that the Jelineks were now skating with one of the big American ice shows themselves, though she wasn't sure which one. Nor did she know their address. She ended up writing to them in care of a big sports arena in Toronto, the Maple Leaf Gardens.

It turned out to be all right. The Jelineks' father answered, promising not only to forward Jana's letter to Otto and Maria who were on tour then, but to make sure that Jana would be contacted by one of the American ice revues in Vienna. And so she was: in less than two months she received a letter of intent from the famous Ice Capades in Los Angeles. They said that they'd put Jana on contract if she could make her way to them.

Jana and Frank promptly applied for a visa to emigrate, but not to the United States. They applied to emigrate to Canada.

They knew perfectly well that Los Angeles was in California. They also knew that California was in the United States. Still, they applied to Canadian immigration because the Jelineks were in Canada, because they knew nobody but the Jelineks in the New World, and because they didn't think that it would make any difference.

It was a mistake, but it was understandable.

Jana, well-travelled as she was, had led the sheltered life of an "official" East European athlete. She had stayed in a lot of hotels, but she'd never had to check into one. She had had to arrange nothing for herself at all. Indeed, they would not have permitted her to make any decisions or arrangements. No globe-trotter has ever been less sophisticated than Jana. As for Frank, he was sophisticated only in the ways of racing circuits.

Like many escapees from communist countries, they also had certain illusions about the capitalist system. Official propaganda always described the West as sheer hell, which made most refugees think of it as pure heaven.

They thought that once they crossed the border into the "free" world they could do, almost literally, whatever they wanted. They didn't realize that just because people in the West were free to leave a country, they were not necessarily "free" to enter another. It didn't even cross their minds that work permits and similar bureaucratic requirements existed everywhere. And, like Europeans in general, they thought of Canada and America as pretty much the same country anyway.

It seemed more important to them at the time not to be entirely alone and friendless in a huge, incomprehensible, English-speaking continent. So, when after about four months they received their Canadian papers, they flew to Toronto. It was only then that they discovered that, Ice Capades or not, no one could work in America without getting American papers first. They found out only in Canada that it might take Jana another four or five months to get a U.S. work permit if she were to get one at all, and that Frank could not hope to get *his* papers to go with her. Not for quite a while, in any event.

It was a discovery that would eventually force Jana to choose between skating and her marriage.

A roast chicken in a garbage can was Frank's first impression of North America. A perfectly good roast chicken with only the legs missing but intact in all other respects. He actually fished it out of the trash bin because he couldn't believe his eyes. Yes, that's what it was, a chicken with the finest meat on it still untouched, that someone had tossed away in Toronto as an object of no value. It was a chicken to remember, and Frank would chuckle and shake his head over it even after twenty years.

It was in the same league as the piles of peaches and lemons in the big supermarket they called Loblaws, where they also had bunches of bananas on display. This was another double-take for Frank. Bananas? It's winter, they have *bananas*? This continent was not to be believed.

The difference between Jana's and Frank's first impressions of the New World manifested itself the minute they landed at Toronto's Malton Airport, as it was then called, a week or two before Christmas in 1965. Mr. Jelinek, as nice a gentleman as one could wish for, had mobilized half the Czech community for their arrival. A young lady named Dana Nohavica was to wait for them at customs and take them straight to the Czechoslovakian Hall for a sports reception.

The only trouble was that when Dana had them paged in the air terminal Jana and Frank didn't respond because they couldn't understand a word of English, not even their own names. They just stood there in the strange, bustling, metal-and-glass arrivals area with their suitcases, wondering what was going to happen to them next. There were telephones everywhere and they had a number, but those alien contraptions were quite different from any telephone in Europe and neither Frank nor Jana could figure out how to use them. Finally an Austrian fellow passenger had to show them how to dial numbers in the New World, which only amused Frank but made Jana feel acutely uncomfortable.

Perhaps the real difference was that Frank would pick up on what was good about North America right away, while Jana would pick up first on what was bad about being an immigrant. It wasn't a question of who was right or wrong. They were both right, only their perspectives were different.

What Frank saw was a country where a couple could get off the plane with their suitcases, then within a few days find a nice furnished flat for $18 a week. Not just anywhere but in a decent area, near Toronto's neat, attractive High Park. Also a job, at $65 a week, at a walking distance from the flat—because one week after their arrival Frank was already working, cooking chocolates at the Nielson factory.

True, the work was hard, loading cocoa beans onto rollers, with the penetrating sickly smell of caramel in his nose all day which made Frank detest chocolates for the rest of his life. But who cared, a job was a job. From day one they could live like kings, have meat any time they wanted, any kind of meat, three times a day. (Frank happened to love meat, which back home not only cost the earth but was pretty scarce; people would line up for a decent cut at the market every Friday.) Here they could eat meat, move into their own place, buy a TV set, it was ridiculous. Frank simply couldn't believe how easy life was in this country.

With Jana it was different. She couldn't help remembering that back home she had been a star, a champion, a national treasure, and here she was a waitress in a greasy spoon waiting for a work permit. Back home she had her entire family: people she loved and, for all she knew, might never see again. Innocent people who might be made to suffer for what she had done.

It never crossed Jana's mind to resent Czechoslovakia, a country that might punish her parents or brother for her decision to defect. (Fortunately, they weren't punished, as it turned out.) Jana was not a rebellious person. All her life she had stayed away from politics. She didn't care about systems: capitalist, communist, they were all the same to her. Some people said one thing, other people said something else, and she was always too busy skating. Anyway, whatever the government at home might have done to anybody else, she knew that it had always been been good to her.

The flat in Canada was nice, true, but they had also had a nice apartment over there. Maybe other people had to wait five, six, ten years for an apartment, but she was representing the country, she would get those privileges right away. She was travelling, never paid a penny, everything was paid for by the government. She had seen half the world by the time she was twenty. Eating meat was fine, but who needed meat three times a day? Who wanted bananas in the winter?

If Jana resented anyone in those first days, it was Frank. She didn't really *resent* him, but facts were facts. He took her away from her well-mapped, well-ordered life where she had skated competitively, then skated in an ice show, then maybe could have coached young figure skaters for the rest of her days.

Now this had all vanished because she had let Frank twist her arm in Ljubljana. Not that it was his fault. She had a choice, she could have said no, so she had only herself to blame. Maybe it wasn't even the *wrong* decision. Jana was far from being blind, she could see that this was a great, wealthy, exciting continent; but if it did turn out to be the wrong decision—well, she had made it herself, but it was Frank who put it into her head. She would never have thought of it on her own.

And now, perhaps, Frank owed her something.

All this was just a feeling, unspoken, barely bubbling under the surface. Unspoken partly because they were too busy trying to figure out this new country and partly because Jana was nothing if not a dutiful, hard-working wife, an almost compulsive saver and nest-builder. She didn't even mind waitressing, it was all right. Not having a job would have been far worse.

Jana complained about nothing, except the language. It was disorienting. It just didn't feel right, not being able to ask people something or to tell them anything.

Frank was different even in this respect. He could somehow talk with his face and his hands. Not that he wasn't interested in learning English, but hell, at the factory people spoke Serbian and Italian and God knows what else, and when he tried going to language school the teacher kicked him out after the first two or three lessons. It was because it was very warm in the classroom. He had been walking in the cold and it was too hot inside, so he kept falling asleep. Didn't mean to be disrespectful to the nice young girl who was trying to teach them English, he just passed out, he couldn't help it.

This was how it started. Then, within a few months, everything turned completely around. Jana received her green card from America and flew straight to Boston to join the Ice Capades on tour.

As far as Jana was concerned, this was it. This was was why they had left. This was the sole reason for going through the trauma in Ljubljana, the wrench from her homeland and language, the heartache of never seeing her parents again. It was all done for her to be a principal skater in a big Western ice show.

Well, now the papers were in her purse and she was flying to Boston.

She had no idea what to expect. There was a contract, but neither she nor Frank could figure out exactly what it said so she didn't even know how much she would be paid. Ask a lawyer or an agent? But the ice revue was an official body. Where Jana came from people didn't question official bodies about anything. She was even less prepared to dispute things in this alien country. She was happy that they were willing to take her.

It was just as well, because everything turned out fine. Not only fine, but beyond Jana's wildest expectations. The show was great, the people were nice, and the money they paid her, whatever it came to, was more than enough. Jana could afford to fly back to Toronto to see Frank right after the first show and tell him all about it.

The next two seasons were the happiest time in Jana's life. She bought herself an entire new wardrobe from her first earnings. Even on top of that she could start saving money. There was enough left to pay for Frank's airfare and hotel whenever he came to visit—which was often enough, once every two months or so when the ice revue was playing the east coast.

The west coast—well, the west coast was a bit too far. When the Ice Capades was touring in the west they couldn't see each other for a while. Still, it was no hardship for Jana. Obviously it would have been better to be with her husband, but that was show business and she could handle it. They could have eight weeks together during the summer months.

Anyway, this was the whole idea. The idea was for Frank and Jana to come to America so that Jana could skate in an ice revue, and now she was skating and enjoying every minute of it. It was so relaxing to glide and act on the ice without the pressure of competition. It was pure bliss. Then there were the TV shows, the interviews—because everybody wanted her to appear on those programs even though she still couldn't speak much English. Her picture was in the newspapers in every town. She was on posters and flyers, even on the Ice Capades' official envelope. It was just like home again, except richer, glossier, and more spectacular.

It was during this period that Jana came close to being glad that she let Frank twist her arm. He had turned out to be right about getting out of Eastern Europe after all. Now Jana could live the life she liked, having to deny herself nothing, and still be able to save for a house. What more could a person want?

Well, a person could feel heart-wrenchingly guilty about her mother and father. For a responsible, dutiful daughter like Jana knowing that she couldn't be there for her parents as they were getting older and needed her the most was a sign of a disloyalty, a failure. This was why she could not be totally at peace about letting Frank talk her into leaving. Except for this, she would have been ready to forgive him, and more importantly, to forgive herself. Because she was happy.

Perhaps not surprisingly, it was Frank who turned out to be unhappy before long.

Of course, in the beginning Frank was happy enough. He was proud of Jana and delighted to see her succeed; that part was easy. He was also glad that he could write to Jana for money—for instance, when he wanted to buy a new car. Not that he ever looked to his wife for support because Frank had always worked: from the fourth day of his arrival in Toronto he had been working like a dog. He wouldn't have known what else to do with himself. Sitting around doing nothing would have driven him out of his mind even if he had been a millionaire (except in that case he would probably have worked mainly on shaving off a few more tenths from his lap-times).

None of this was a problem, but what Frank found difficult to handle was their separation. At least, as it appeared to Jana, he found it difficult to handle it with *integrity*.

Jana could understand that this was a bigger problem for some people than for others. For her, as it happened, it was no problem at all. Sure, sex and companionship were great; if you loved a person you obviously wanted to be with him in every conceivable way. But if he didn't happen to be around, well, you just did without.

All nice girls would learn to do without under such circumstances, and it was a pity that so many men could not learn to do the same thing. As Jana saw the matter, it was just a question of integrity; there was no other word for it.

Frank could probably have used other words, but words were never his thing, so before very long it boiled down to him starting to push Jana to quit the show. Quit the show, come home, and give up skating. If Jana wanted integrity it was all very well, but Frank was not made of wood and what he wanted was his wife.

Jana was shocked. This was the first time, the very first time, that the thought had even entered her mind that when Frank talked her into leaving Czechoslovakia it wasn't just for her sake. Frank must have also been thinking about himself. He did not make her go through all this trauma just for the sake of her own career, but *he* also wanted certain things.

It would be easy to say that this ought not to have come as a surprise to Jana; the fact was that it did. Maybe she ought to have taken it for granted that all people have desires, Frank included, but it had never occurred to her that Frank might want anything else out of life than to make her the star of an ice revue. Should she have thought of it? Why, Frank never brought up a single reason—any reason at all—for abandoning their home except the ice show. The very show that he was now asking her to quit.

Quit? But then why had they come here?

Jana was shocked, hurt, and puzzled. Of course, she had always known that Frank was enchanted with capitalist countries. He liked them because they made those Mercedeses and God knows what other types of cars. Frank liked machines, so wherever they made nicer machines that was where Frank preferred to live. All right, but he hadn't told Jana that they should leave their own country for the sake of nicer cars: he had said that it was all for Jana's skating.

Maybe Frank had been lying. Maybe he wanted to leave for his own purposes more than for her happiness.

Also, Frank had always been jealous. From day one. Not that he had the slightest reason—he probably didn't even *fancy* that he had a reason—but being jealous was a kind of reflex with Frank. He had never wanted Jana to give up skating before, though, so she just put up with it. Perhaps (and this was crossing Jana's mind for the first time) Frank might have been jealous of her travelling all those years, too. He had never said so, but maybe he was jealous of her flying to exotic places all over the world while he couldn't go anywhere except for that one trip to the Olympics. Maybe in Ljubljana he had twisted her arm so that he could finally travel himself.

And now that Frank was in Toronto, cooped up in one place again while she was going with the show all over America, maybe he just wanted to spoil her fun.

Jana felt pained and disappointed by her discovery—because she considered the thoughts crossing her mind as a "discovery" at the time—and she decided to get a divorce. Her first two seasons were drawing to a close, and Ice Capades was offering her a new contract for two more seasons. A new contract in a principal role and for even more money. Jana was almost sure that this would be followed by yet another offer when the new contract expired. She liked the revue and they liked her, and it was possible for people to skate in an ice show for seven years or even longer.

Jana made up her mind to get a divorce because the contract

was there for her to sign and Frank kept pressing her to quit and she had to make a choice.

The choice that Jana *meant* to make was a divorce. However, people don't always do what they mean to do, and the choice that Jana made in the end was something else. Her choice was to refuse the Ice Capades' contract and to stay with her husband. In September 1967, she said good-bye to the show and went back to Canada. It was impossible get a job in Toronto at the time, so she accepted a coaching position in North Bay, a small industrial city in northern Ontario.

That was Jana's choice, and she could never really understand why she had made it.

Perhaps it was because Frank had said to her that they had no one but each other in this new country. Or perhaps because (as she occasionally accused herself afterwards) she was a weakling. Perhaps because, having been disloyal (as she thought of it) to her parents and to her country, she did not want to be disloyal to her marriage vows as well.

But perhaps Jana chose as she did simply because when all was said and done she loved Frank more than she loved skating. It probably *was* the real reason, but contrary to popular belief, love does not cure all. Love may have made Jana make her choice, but it did not make her choice any easier. In fact, it continued to hurt and rankle her for the next twenty years.

11

Morning Practice

The ritual is always the same.

On race or practice day, around eight o'clock in the morning, as the smell of sausages and bacon wafts over the camp sites, the drivers begin to gather at some designated spot near the mock grid on pit row. A few come from the paddocks riding bicycles or tiny pit bikes, but most just saunter or limp along on foot. One or two may hobble on crutches. (The crutch-and-cast crowd are likely to be young guys: when you're young, you don't want to miss a single ride.)

No one ever hurries, even though the track loudspeaker has been issuing plaintive calls for the last fifteen minutes. To be seen hurrying would be against track etiquette. Once in a while a rookie may break into a trot, but then he checks himself and plods along with the rest.

The racers come singly or in little groups, chatting, yawning, sometimes sipping coffee from thermos tops, bundled up in sweaters and ski jackets in early spring or wearing shorts and T-shirts in the summer. Some put on their lucky hats or toques.

A few riders may already be wearing their one-piece racing leathers, but seldom zipped up all the way. The top part is either tied around the waist or is allowed to dangle carelessly to the ground. This is the dress code at the drivers' meeting before morning practice.

Every event, from club races to Grands Prix, begins with a compulsory drivers' meeting. It is there that the referee of the course, the starter, the chief marshall, or their helpers issue their order of battle to competitors. If a racing event is spread over several days there is still a meeting each morning before the first bike goes on the track. There must be a meeting, whether the officials have any exciting news for the drivers or not.

Veterans use those parts of the ritual which never vary to catch up on their sleep. For instance, while the referee is "going through the flags". With minor variations track-side signals tend to be the same at every circuit all over the world.

The stationary yellow flag means caution and indicates some incident just off the track. When the same flag is waved it means extreme caution: maintain racing speed but do not attempt to pass. A waved flag signals a recent incident right on the racing line.

A steady white flag means a service vehicle somewhere on the circuit, usually an ambulance. A white flag waved means service vehicle straight ahead, no passing. People who are caught passing on waved flags are penalized a lap or a position.

The black flag, pointed at a rider or displayed along with his racing number, tells him to complete the lap, come in and see the referee. If he's smart, he'll check to see if both his wheels are still firmly attached to his machine. Being "black-flagged" means that either the driver or his vehicle is out of bounds. It doesn't just mean that somebody's zipper is undone; either he is riding dangerously or his bike is shedding hardware or dripping fluid. Getting the black flag is a bit like being called into the principal's office, and drivers hate it.

The red flag means stop. The race or practice is being halted. Stop, but preferably (the referee adds, for the benefit of rookies) don't stop *dead* in your tracks. Just stop racing, pull off the course or coast slowly to the nearest marshall-station and wait for instructions.

Nobody wants to see the red flag. It usually means the weather turning hopeless, such as thick fog rolling into some valley on the course. Or it means a serious, maybe multiple accident.

The striped yellow-and-red "oil" flag signals obstacles or slippery conditions—usually straw, oil, sand, or other debris on the track. Corner workers may point the striped flag at the danger spot. They point it at the spot itself and (for the benefit of rookies again) *not* at the safe part of the course. People are inclined to go wherever they see someone point, and some young riders have come to grief by driving into the very stuff from which the marshalls have been trying to warn them away.

A green flag means a clear course. Crossed and furled blue and green flags signal the halfway point of a race. A blue flag with a diagonal white stripe means go for it: you're on the last lap. The position you pick up (or lose) now will be your final position in the race. A blue "passing" flag is sometimes waved at a competitor holding up a faster driver, though it's rarely used at short circuit motorcycle events.

The chequered flag, of course, means that the race is over. Take a cool-off lap (on most tracks) but stop racing. This injunction is necessary because the adrenalin is still flowing and riders have been known to go on dicing, just as some boxers keep punching after the bell. So stop racing—but don't start a strip-tease. Once in a while riders go into a zombie-like state at the end of a race and release their tension by unbuckling helmets, unzipping leathers and shedding gloves during the cool-off lap.

The most important signal is the starter's. It may be a white flag, raised high with both hands for "in gear" then dropped for "go". It may be used in conjunction with a board displaying first a three-minute and then a two-minute sign, finally turned sideways a few seconds before the flag. On some tracks ordinary traffic lights are used, red, yellow, and green, with the starter mounting a platform and holding the switch high before hiding it behind his back. The starter must hide the switch, or else people in the front rows would watch his hand instead of the light to get a split-second jump on the field.

Good starts are important but not easy to achieve. You may keep gunning the engine to keep the revs up, which means rotating the twistgrip on and off, but if the flag happens to drop just as you've closed the throttle the thing might bog on you. If you try to anticipate the starter by slipping the machine into gear too early you may fry the clutch. If you're too late with the gear you may get stuck at the start, or worse, end up with a bike rammed up your backside. When the flag drops many guys just dump that clutch and go, and too bad if you happen to be in the way.

Dumping the clutch may result in a big wheelie. Spectators love a wheelie but it loses you time. The bike dissipates the energy that should be propelling it forward by trying to climb up the chain, and while the front wheel is pawing the air the machine is both slower and less stable. You'll probably have to roll off a bit to let the wheel settle back on the ground, and meanwhile a less spectacular starter next to you will have moved twenty yards ahead. In an ideal start the front wheel would be lofted just enough to skim the pavement and settle back under full power as you keep shifting up through the gears—but who has ever had an ideal start?

Perfect starts are even harder to achieve when competitors must switch off their engines after the warm-up lap, then bump-start their machines at the flag in the classic fashion: at Grands Prix in Europe, for instance. (Luckily motorcycle riders rarely have to do a Le Mans-style foot race before they ever get to their vehicles.)

Bump-starting requires both drivers and machines to be in top tune. It entails running with a bike anywhere from two to six steps, then jumping on it sideways while releasing the clutch and praying that the engine will fire.

Before any of this happens, though, both bikes and riders must be "teched in". Scrutineers must look over the bike and the protective gear of the racers, pasting their stickers of approval on both men and machines.

Racers and machines must conform to the regulations of sanctioning bodies—FIM internationally, AMA (American Motorcyclist Association) in the United States, along with whichever affiliated national federation or club is organizing the event—as well as to the specifications of their class. All this is governed by complex rules. Since everyone is looking for the winning edge, some honestly and some any way they can get it, adherence to regulations is monitored pretty closely.

Next, the scrutineers attempt to make sure that the machines are safe. In a sense this is an impossible task. Short of taking apart every bike and X-raying all critical parts no one can tell whether or not they may fail under the stress of racing.

Still, the scrutineers try. For instance, they look over fastenings that could work loose and ought to be drilled and lock-wired. They insist on at least drain plugs and certain types of oil-filter housings being wired for safety. They glance at frames for visible cracks and may test critical pivot-points such as steering heads. They check breather-type fluid lines, which (unless equipped with one-way valves) must drain into catch cans. Some circuits adopt the Grand Prix rule of bikes being able to lean fifty degrees from the vertical before anything touches the ground. In our environmentally-conscious times even rules restricting noise levels have been introduced. None of which have prevented machines from failing on the track, hurting or killing their riders and others.

Bikes can fail in innumerable ways, but some of the worst and most frequent failures in the past have involved engine seizure, especially on two-stroke motors. It was Renzo Pasolini's seized Aermacchi two-stroke at Monza that killed both him and world champion Jarno Saarinen. A few years earlier, in 1969, a two-stroke Jawa seizing at East Germany's Sachsenring circuit caused the death of former 125 cc world champion Bill Ivy. Of course, Pasolini, Saarinen, and Ivy were simply among the most stellar names in international racing who fell victim to seized engines; there have been countless others, though luckily not always with fatal results. Other fairly common and potentially deadly mechanical failures have included blown engines dumping oil on tires and race tracks; carburetor sliders sticking open; front brakes failing; primary or final drive-chains breaking; and, less frequently, wheel spokes collapsing or tires coming off their rims.

There isn't a great deal a racer can do about some types of mechanical failure. You can triple check everything—in fact, you're a fool if you don't—and then pray that it doesn't happen to you or to the guy next to you. Prayer is as important as triple-checking. You realize this when the first chain that fails on you turns out to be a spanking new, fresh-out-of-the-box item with which you've prudently replaced your old, worn-out chain—unlike your careless buddy who ran his old chain with no trouble throughout the season.

That, too, is racing.

The rider's own equipment is essentially cowhide of approved thickness, preferably of one-piece construction and permitting no part of the rider's skin to show when seated on the bike, except for an inch or so at the neck. Since the late 1970s some people have lined their racing leathers with exotic shock-resistant materials such as Kevlar and have worn hard plastic knee-pucks and even back braces or protectors, all unknown in the old days and not mandatory even today. Leather gloves and boots are required items, and often they are studded with little metal discs where they might come into contact with the pavement.

All this is excellent protection against abrasive types of injuries, such as the infamous "road rash". It offers only minimal protection against impact-induced injuries. Cowhide or even plastic can't shield you much against hitting a solid object at speed. The only solution is to remove all solid objects from around the track, or at least wrap them in straw or rubber—

but there's one solid object you can neither remove nor wrap in anything, and that is the ground.

If you fall in such a way as to slide along the ground, whether on the asphalt or on the grass of the track-side run-off, your leathers will protect you well enough. You may simply skate along on some part of your anatomy until you stop. This is what often happens when a bike is "lowsided" or dropped in the direction of its initial lean or skid. It is possible to escape very high speed, 100-plus mph lowside spills without significant injury. In fact, unless you start tumbling, you're more likely to be uninjured than not.

When, however, a bike is "highsided"—generally because its sliding tires encounter sudden traction or resistance—it may spit off the rider and fling him into the ground. Sometimes the bike may land on top of him. A highside spill is more likely to be followed by a tumbling slide, resulting in repeated minor collisions with Mother Earth. An accident of this type, even at lower speeds and without any subsequent impact with track-side objects or other bikes on the circuit, is hard to escape without injury. The most frequent kinds include fractured collarbones, ankles, wrists, or toes. Also popular are cracked ribs, dislocated limbs, and separated shoulders.

Racers, just like skiers or football players, take such injuries in their stride. They are the price of admission to a way of life. People do everything to avoid them, but seldom become disheartened when they can't.

Back, face, and head injuries fall into a different category. Luckily they are much rarer. They have become rarer partly as a result of safer tracks, and partly because of helmets, which have evolved from little bean-shaped leather-and-plastic bowls perched high on top of riders' heads to modern, full-face, space-age affairs. A helmet of approved industrial standard for racing—currently SNELL 85 in North America—can withstand considerable impact. It can, in most cases, protect the rider against any concussive forces he is likely to encounter on well-designed race tracks.

Even the best helmet can't protect against all impacts,

though, or against obstacles on circuits of poor design. Or against fatal injuries to other parts of a rider's body. (They certainly can't protect street riders who hit a tractor-trailer head on, for example.)

In the course of withstanding an impact a modern helmet self-destructs. This is the nature of its design. Theoretically it should be replaced even when it's just dropped on the ground. At costs ranging from a minimum of $300 to over $800, this is bad news. The good news for some is that top racers may get helmets for nothing from manufacturers as promotional items.

With registration completed, entry fees paid, waivers signed, scrutineering stickers attached to fenders and helmets, the riders take to the track for morning practice. They usually embark on their ten-to-twenty minute sessions class by class: tiddlers with tiddlers, big boys with big boys, to prevent machines or riders of vastly different speeds or abilities from mingling on the track.

"Vastly" is a relative term. On a short circuit of a mile to 1.5 miles the difference between the slowest and fastest rider within a given class is rarely more than seven or eight seconds a lap. Often it is less than five seconds. That, of course, is sufficient for race leaders to lap some backmarkers by the end of an eight-to-twelve-lap sprint. On Grand Prix-length tracks the difference between fast and slow guys can be much greater—which shows whenever regional or club races are held on longer tracks—but at international levels riders and machines are more evenly matched.

Riders tend to take it easy on the first lap of practice to warm up their tires. (Before an actual race they've already had a warm-up lap.) No one takes it *too* easy, though, or else the tires don't get warm enough. Racing skins at their best temperature for adhesion are quite warm to the touch and look "scuffed in", shedding tiny flecks of rubber. When they get too hot and "greasy" they lose some adhesion again.

Real practice begins after the first "flying lap" past the start/finish line. This is when racers buckle down to serious business. They look for the best line through corners, feel for the limits of traction, experiment with different peeling-off points or gears through various hairpins and sweepers. They may also pick out reference points to begin and end their braking. (Picking bikini-clad girls in the crowd can be risky in case they fidget or move elsewhere.)

People who are new to a track will try to learn the course itself, but even veterans must learn the course as it is on a particular day. There may be new bumps or not-quite-healed oil spills on a rider's usual line. The weather, even short of dramatic things like rain, can make a big difference. Though on some tracks, especially in the United States, races are called off in the rain, people will race on very hot or chilly days even in safety-conscious America. Temperature influences the performance of tarmac, tires, and engines. Jettings may have to be adjusted. A braking marker that has served you well in moderate weather will put your tires over the limit on freezing or sizzling days.

In morning practice people first try to sort out the engine, the suspension, the tires, and the course. Then (or, better still, at the same time) they start sorting themselves out. Racing is primarily a battle with the clock. Before you can think about passing anyone you first have to overcome the problem of being slower than he is. Once in a while it is possible to steal a race from someone who is (marginally) faster than you, but on the whole you can't race with anybody unless you are in the same ballpark in terms of lap-times.

The most important component of any racing bike sits in the saddle. To find "The Edge" that everyone talks about, you first must look into yourself. People tinker with their machines and rightly so because machines can make a difference, but anybody more concerned with his machine than with his riding can tune himself right out of a race. True, a world-class rider can't give away even a single horsepower to another world-class rider, but he can give away thirty ponies to everybody else. For a difference in your bike's performance you pay in dollars and measure the return in pennies. Investing pennies in your own performance can profit you in dollars.

Racing is a bit like learning a computer game. At first everything is a blur, but the faster you get the more things slow down around you. Your body may be hurtling through space at 140 mph but that's nothing when your mind can move with the speed of light. Things seem to whiz by only as long as your mind lags behind. Once your brain has learned to be just fifty yards ahead of your body the real world is reduced to slow motion.

You have peeled off at the entry to a turn, leaning to the limit and moving so fast that spectators can hardly follow you with their eyes, but to you it has become an interminable wait. You are all but drumming with your fingers on the twistgrip in impatience, waiting for the moment when you can pick up the bike a little, turn on the throttle and begin the drive out of the corner. That long middle part of a turn when you just sit there doing nothing—in reality maybe a second—can seem endless during a race.

This, incidentally, is one reason why people sometimes begin the drive through the exit of a turn just a fraction earlier or harder than possible, while they are still leaned over to the limit. It is a good way to break loose the rear wheel and find yourself sitting on your ass.

Once a rider has made his peace with the clock and is moving as fast as the first guy he can reasonably expect to catch—which may still only be somebody in twentieth place—he can begin to practise his racecraft. Passing someone on the track is an experience of almost spiritual dimensions, and the rarer it is in a racer's career the more satisfying it can be.

The simplest pass is on the straightaway. It's achieved either by having more horsepower than someone else or by getting a better drive out of a corner. Engines take some time to rev up to speed, so if you can turn on the wick a bit earlier than your competitor does you may catch him on the straight even if he's riding a bike of similar performance. But a lot of passing is also done in corners, either at their entrance "on the brakes" or at their exit "on the gas". (It is possible to go around someone in the middle of a corner, usually on the outside and provided that you are quite a bit faster, but trying to "close the door"* on somebody near the middle of a turn leaves next to no margin for error.)

Probably the easiest way to attempt a pass is by outbraking someone at the entrance of a turn on the inside. It is by no means the safest or the most elegant method, but it is the least scary to initiate. (It may get quite scary a split second later if you overcook, but that's a different story.) If you can stick a wheel inside a competitor before he peels off for his turn you may "steal his line". You'll be quite tight and slow through the corner yourself, but perhaps he'll be forced to tag along behind you. The risk is that he may re-pass you at the exit on your inside when the line you've chosen forces you to go wide.

It is also possible to outbrake people on the outside at the entrance of a turn. Then you can close the door on them, or, less happily, get T-boned by someone who doesn't take kindly to having doors closed in his face. This can also happen when you've been on the outside to begin with, but some charmer sneaks up inside you, so the risk is probably the same.

To pass somebody on the gas at the exit is safer and more stylish. It is also more efficient. It probably requires you to be a faster rider, though, with more momentum through the entire turn. In any event, racing is not a sport that awards points for style. There is nothing wrong with any method of passing as long as it works.

The inside is considered safer in close dicing. If the guy on the outside loses it he will bounce harmlessly away—harmlessly to you, that is—but if somebody on the inside loses it he will probably take you down with him. However, this is less of a consideration than being in the right position for whatever you hope to do next, which is often another turn. Passing someone is of no earthly use if it messes up your line for a

*The act of passing a competitor on the outside and immediately tightening the radius of the turn ahead of him. Sometimes the term is also used to describe any completed pass that results in the passer occupying the competitor's line.

corner that is just coming up. He'll re-pass you, grin, then fade into the distance.

If it's your lucky day and all goes well during morning practice, your turns will soon flow into one another. You'll catch the groove and become a note in the harmony of the line. Instead of fighting gravity and inertia, you'll be part of the physical forces of the universe. You'll find yourself skating effortlessly on the fine edge of friction, precession* and momentum. For a few glorious minutes you will translate your mass into energy, square your own velocity and become the Einstein principle itself, the continuum in time and space. (Well, maybe not $E = mc^2$ because your own velocity is hardly the speed of light, but $E = mv^2$, where v is your own speed at the start, is nothing to sneeze at.)

Then, all too soon, you'll catch the chequered flag from a corner of your eye, signalling the end of practice. You'll back off the throttle a bit, sit up straighter, maybe lift the visor of your helmet as you go around your cool-off lap. After flying it is always a bit of a bother to come down to earth.

You'll stick out a hand or a leg to signal at the exit before pit row and coast to your spot in the paddock. When you have put the bike on the stand you'll peel off the top of your racing leathers. You'll wipe your hands, and as a worthy representative of the time-space continuum treat yourself to a bottle of pop.

Frank was aching for his first morning practice in North America. He could almost smell it and taste it. For him it wouldn't just be a practice for a race but for an entire new way of life. However, as he soon discovered, there were a few obstacles.

First, unlike Jana, he had never been an international figure in his sport. Czechoslovakian SVAZARM championships

*The motion of a spinning body, such as a gyroscope, sweeping out of its cone. It is this force that requires motorcycles to be "countersteered", i.e. pushed into a turn by first steering them slightly in the opposite direction.

meant little in the U.S. or Canada. The only people who had ever heard of Frank were a handful of international riders, like Mike Duff, Gary Hocking, or Jim Redman, against whom he happened to race in Brno. Even they might not have been bowled over by the eager young Czech since they could generally beat him.

Sadly, Gary Hocking would no longer be a reference for Frank in any event. After winning the 350 and 500 cc world championships in 1961, the handsome Welsh-born Rhodesian of a different time warp started the 1962 season with high hopes. The first race was the Junior TT at the Isle of Man. Hocking's MV Agusta came second to Mike Hailwood's, but it was in this race that Hocking lost a very close friend. The Australian rider Tom Phillis had been in third place on his Honda, lustily chasing Hocking's and Hailwood's Italian machines, when he was killed hitting a wall at Laurel Bank.

Hocking was deeply affected by his friend's death. He still rode in the Senior TT the next day and won it, but then immediately announced his retirement from motorcycle racing. Cars seemed a much saner choice, so Hocking returned to Rhodesia to contest Formula One on four wheels. In December the same year he died when his car crashed during practice for the Natal Grand Prix.

Hocking had been twenty-one when he first joined the "Continental Circus", as riders called the European series of Grands Prix, and twenty-four when he became world champion. Had he lived, Gary Hocking would have been twenty-eight in 1965 when Frank arrived in Canada.

By then Frank himself was twenty-nine. He was not too old for international racing perhaps, especially in those days, but he was a bit too old to start.

Twenty-five-year-old Mike Hailwood—"Mike the Bike", eventually twelve-time winner of the Isle of Man TT, multiple world champion, recipient of the George Medal and Member of the Order of the British Empire—was already the winner of four world titles for his MV Agusta team in 1965. It was the same year that a young Italian named Giacomo Agostini, later winner of fifteen world titles, first came on the international scene, also as a rider for Count Corrada Agusta. By 1965 top riders of Frank's own age, like New Zealander Hugh Anderson, had often won their world championships and would soon retire from competition, as Anderson did in 1966.

"King Kenny" Roberts, America's future triple-world champion-to-be in 1978, '79, and '80, had not yet tried racing on a road circuit, though he might have had a race or two under his belt on the dirt track. Eventual triple world champion Freddie Spencer was only three years old in the United States. In another three years, at the age of six, Spencer would start racing already, and win his first world title in 1983 at the age of twenty-one and both the 500 cc and 250 cc world championships in 1985. Britain's double world champion of the 1970s, Barry Sheene, would not start competing at all until 1968, three years after Frank's arrival in the West.

The only rider from Eastern Europe who defected and then made the world scene in those years was Ernst Degner. In 1961 the East German MV factory rider was points leader in the 125 cc world championship class when he surprised the racing world by defecting. When he couldn't immediately get a competitive ride in the West, Degner lost the world title to Hocking's friend Tom Phillis, but eventually started racing 50 cc machines for the emerging Japanese Suzuki factory. He won a world title for Suzuki in that class next year, made a comeback after burning himself badly in a crash in Japan, and in 1965 won at Daytona for his Oriental manufacturer.

Ernst Degner was Frank's senior by about five years. When he won at the famous Florida track he was already thirty-four. If Degner could do it, there was no reason why Frank could not give it a try.

However, there was another problem. In addition to having to start from scratch at a relatively late age, Frank made the astounding discovery that for his particular sport he had chosen the wrong country once again. He chose the wrong continent, in fact.

Americans and Canadians are, of course, creatures of a high

automotive culture. They drive their big cars more frequently, over longer distances, from an earlier age and in larger numbers than any other people on earth. However, by the mid-sixties they rarely rode motorcycles and they certainly did not race them much on road circuits.

They didn't even race cars too often on road circuits. Big-time racing in America was Indianapolis-style oval speedways, dirt tracks, and maybe motocross or enduro. Road racing as a spectator sport, with negligible exceptions, was pretty much at the bottom of the list. There were some fine circuits like New York's Watkins Glen, capable of accommodating Grand Prix style racing, but the sport was thought to have only limited spectator appeal.

Motorcyclists in America during that period were usually called "bikers" and had an unsavoury, near-criminal reputation. Some congregated in gangs, and while only a minority engaged in anything worse than quirky behaviour, the few who did were often bestial enough to put their stamp on the rest. Even apart from these "one percenters", as they called themselves, many American riders at the time had more interest in some juvenile, mock-image of speed and courage than in its substance. They played the part of he-men or daredevils, trying to look a study in defiance and machismo while rumbling slowly around on huge Harley-based "choppers" with skinny front wheels, ludicrously extended forks, and high "apehanger" handle-bars.

Frank could hardly believe his eyes when he first looked at them. Motorcycles? A tractor could have passed them going around a corner.

Admittedly, a few choppers were carefully crafted and rather beautiful expressions of industrial art. They were show-bikes, hand-painted, gold-plated and polished to gleaming perfection. Their riders included some of the sweetest and most decent human beings, solid, old-fashioned craftsmen, who rebelled against nothing but the cookie-cutter spirit of assembly-line existence.

Even so, their creations were useless for the one thing that interested Frank, namely racing. He didn't come all this way to stand around gawking in some chrome gallery of modern art.

Many die-hard enthusiasts or collectors from an earlier era continued riding their Harley-Davidson touring bikes during this period—indeed, some of them rode vintage machines that were no longer manufactured, mainly Indians or other American marques—and there were also a handful of connoisseurs of classic or modern German, Italian, and British machinery. You could even, as the ads put it, "meet some of the nicest people" riding on a Honda scooter. The little toys were just beginning to establish themselves in what later proved to be an Oriental revolution in two-wheeled transportation.

But scooters had nothing to do with racing either. At a time when circuits like Brands Hatch or Donington Park in England could bring in 50,000 spectators for national motorcycle events—and any European Grand Prix could count on 100,000 to 300,000—even the biggest American motorcycle classics like Daytona were lucky to draw 25,000 to 30,000 people. Road racing in America was a hobby for hard-core enthusiasts and their followers rather than a major spectator sport.

Professionals on motorcycles were doing it on the mile and half-mile ovals of speedways, which continued to flourish. It was as American as apple pie—and as the Milwaukee iron that dominated its scoreboard. The dirt track was home turf for Harley-Davidson. Eventually some of the world's best road racers would emerge from the ranks of dirt trackers, but that was then still ten or fifteen years in the future. Speedway racing—sliding bikes sideways along the loosely-packed dirt, weird bikes with powerful engines but with no gearshifts or brakes—was probably thrilling to watch, but it was something that Frank neither knew nor cared much about.

Canada was actually in a better position for road racing. The sport had a following of sorts, and also a few fairly good circuits, including what was then one of the finest tracks in North America for Grand Prix-style racing: Mosport, near Toronto.

Frank knew about Mosport, but how would he get to race there? He arrived in this country with nothing. Who would sponsor him, who would give him a ride?

The subject of racing hardly came up between Frank and Jana during the first twelve months after their escape. They had their hands full simply coping. They had to get documents, money to feed themselves, and a place in which to sleep. Jana tacitly assumed that her husband would stop riding competitively in the West. He'd get a job, and maybe race as a hobby once they were comfortably settled. After all, that would probably take a few years. By then Frank would be well into his thirties, a man with family responsibilites and not a competitor in any real sense of the word.

Frank made no such assumption. He took it for granted, just as tacitly as Jana did, that he *would* continue racing. "How" or "when" would take care of itself. He'd race again, for what else was there for him to do?

If the years between late 1965 and early 1967 were among the best in Jana's life, they were close to being the worst in Frank's. If the New World was working out for Jana splendidly, it was hardly working out for Frank at all. It was ironic: Frank was really excited by America with all its gadgets and freedoms and felt very little nostaligia for the old country, while Jana's feelings were pretty much the reverse. Yet Jana was making it big as an ice revue star and Frank was cooking chocolates in a factory.

True, he wasn't cooking chocolates for long. After about six months a Czech-German car salesman got Frank a job as a mechanic at a Volkswagen dealership. It was a hundred percent better than shovelling cocoa beans, as well as more lucrative, though just as far from racing. Motorcycles hardly came up in Frank's life at all until a meeting with Mike Duff in 1966.

Duff had been a factory rider for Yamaha, well up in the world championship standings throughout 1965. He was the most successful international racer Canada had produced until then, or perhaps ever. Certainly no one came close to

Duff until the appearance of Yvon "SuperFrog" DuHamel on the racing scene some years later.

After their escape from Yugoslavia Frank had bumped into Duff at the home of a mutual Austrian friend. They renewed their acquaintance, played some soccer, then said good-bye. At that time Frank didn't even know that he would be coming to Canada. As for Duff, he was off to his next race on the Grand Prix circuit.

The way things turned out, Duff made it as far as Japan, where he crashed badly in practice. Frank heard about it only in Toronto where the Canadian rider was flown back for hospital treatment. It was while Duff was recuperating at his parents' place in the city that Frank visited him first. Someone had made a short film of Duff's crash and the subsequent operation and Frank could hardly watch it, it was so bad. The crash, that is, not the movie. The film showed how they were fixing Duff's leg, fitting him with an artificial hip or whatever.

By then Frank could speak some English, a little more than before anyway, so he and Duff became closer. Mike Duff, as Frank saw him, was a quiet guy, thoughtful, friendly, a bit like a professor. Very calm, and very fast on the track. Nobody gets to where Duff was by being anything but fast.

In any event, Frank was aching to race and Duff had contacts, so he got Frank his first real motorcycle job at The Checkered Flag, Wally Esden's Yamaha dealership in Toronto. By then Japanese bikes were gradually establishing themselves on the world's racetracks, not only in the tiddler categories of 50 and 125 cc's, but in 250, 350, and even the 500 cc premier class. They often had astounding motors (with some rough edges) and mediocre frames. But they were real motorcycles. Some had racing potential even in street trim and Frank's eyes lit up as soon as he could look at them more closely.

Esden had talked with Mike Manley, a well-known racer and Honda distributor, and the end result was that Frank found himself one day aboard a 450 cc Honda on the old Harewood circuit. Harewood was a former airstrip in southern Ontario, used for training pilots during the war. On a sliding scale it was maybe a three out of ten as racetracks go, but for Frank it was a chance to impress everybody in this new country. He could show them that even if he couldn't speak much English, he could ride.

At the end of the day everybody was duly impressed. Frank's boss, Wally Esden, came to the demonstration and decided to get a Yamaha TD-1B, a 250 cc two-stroke, for his new employee. Doing well in production classes—street motorcycles with little or no modification—was important for dealers. Racing wins could be translated into sales for the still novel Oriental brands. Maybe this strange Czech fellow could clean up with a competitive bike.

Frank did seem a little strange to many people. He really started picking up English only at Wally Esden's shop: a colloquial racetrack English, laced with technical slang and delivered rapidly with a syntax and pronunciation that were purely Czech. Only someone who had both English and Czech (and knew something about racing) could fully understand Frank's language, but he spoke it enthusiastically and without the slightest inhibition.

Frank seemed restless, almost like an autistic child, except when sitting on the starting grid. On the grid he was like most high priests of speed: calm and serene. Otherwise he would either move about like quicksilver or fall sound asleep, with little in between.

In those days Frank didn't always know what was happening around him and he didn't much care. He wasn't even sure what championship series he was competing in. He wasn't familiar with most of the local racers. He didn't know whether he was dicing with Americans or Quebecers. He had no idea how to spell people's names, important people, like promoters or backers. "I don't understand anyway what these guys are talking about," he'd explain later with a shrug. "I just go race."

The important thing was, Frank now had a sponsor. He also had a job and a German Shepherd puppy. He had a clean, decent room and as much to eat as he wanted. He wished he had his wife, too, but she was skating somewhere in California or in Washington State.

Frank missed Jana, but at least on weekends he would again be sauntering along pit row, sipping coffee, wearing his lucky straw hat, and listening to the track announcer's plaintive voice coming through the loudspeaker like a muezzin's cry from a minaret calling the faithful to morning prayer. Nothing else really mattered. It was the spring of 1967 and Frank was thirty-one years old.

12

CORNER 3

He'd fly through the air with the greatest of ease,
This handsome young man on the flying trapeze. . .
GEORGE LEYBOURNE

When a rookie goes to Mosport, some old-timer is likely to put him at ease at the drivers' meeting before morning practice:

"Never raced here before, eh? Oh, you'll love it" is the way most veterans begin. "Just watch out for Corner Two. Boy, you go over that hill the first time, you'll mess your pants. I did, anyway. Feels like you're falling off the end of the world."

Drawn as an outline, Mosport looks like an amoeba that's about to split. The deepest point of the split is Corner 2, a 114.5 degree, 500-foot radius left-hander. It apexes exactly half a mile from the start/finish line.

Corner 2 is a "fast" corner; at racing speeds most people try to get around it flat out, one or two gears down from top gear. Speeds will vary depending on the class of machinery used, but the range might be from about 70 mph for vintage motorcycles equipped with street rubber to about 120 mph for multi-cylinder Superbikes with six-inch slicks. (Incidentally, racers rarely note "speed" as such, and almost never talk about it in miles or kilometres per hour. They measure it in terms of engine revs, lap-times, or maybe by the heating or scuffing of their tires.)

Like any corner, Corner 2 exists both as a physical entity and as an idea in a racer's mind. It is scary because it creates a scary impression. A blind, downhill turn, it feels as if it teeters at the edge of a cliff. In reality it doesn't, but it's the idea that counts. Corner 2 has probably hurt more people than any other corner at Mosport, and some may have been hurt by the idea alone. People lost it in Corner 2 when there seemed to be no reason for them to lose it.

Mosport, about sixty miles east of Toronto, is one of North America's major road-racing tracks. It is spoken of as a "technical" track. Essentially that means that cool, calculating, experienced racers will do better on it than racers who are merely aggressive. To some extent this is true of all race courses, but flat, predictable "go-cart" tracks may reward sheer guts to a greater degree. Hilly courses with blind turns put a premium on expertise. Your heart unaided by your mind may win you a race at Shannonville's Nelson short circuit, but the same combination at Mosport will just increase your heart rate.

During the 1960s some of the greatest drivers of the International Grand Prix circuit used to pilot their machines around Mosport's 2.459 miles of bumpy surface. Stirling Moss, probably the best-known driver of the period, gave his name to the track's slowest, most "technical" turn. Corner 5/B, a 90-degree, 34-foot radius right-hander, is still referred to as "Moss' Corner".

Moss' Corner is roughly at the halfway mark of the course. The track descends to it through Turns 1, 2, 3, and 4 (also known as The Chute) which are all "fast" downhill corners. Some top racers may never touch their brakes until the entrance of Corner 5/A, a 99-degree, 160-foot radius right-hander, 166 feet below the highest elevation of the circuit. There everyone brakes for dear life because Corner 5/A cannot be negotiated until a racer scrubs off about 40 to 60 mph from the speed at which he has been proceeding through The Chute. Since he has only about 120 feet (or at most two seconds) in which to do it, he'll squeeze his brakes until his eyeballs pop out. Front discs wilt in Corner 5/A faster than orchids in a drought.

Moss' Corner is just past the deepest point of the track, at an elevation of about 960 feet, from which the road climbs in a series of undulations to its summit of 1,103 feet at the start/finish line. The small hilltops that mark the end of each descending-ascending jog obscure the track beyond, making the entrance blind into many of the turns.

Two hilltops stand out, literally and figuratively, on the track as well as in the minds of many racers. The first is before the infamous Corner 2. The second comes near the end of the back straight, called the Mario Andretti Straightaway, on which racers reach their top speeds, anywhere from 110 to 200 mph. Riders feel airborne as they're cresting this hill, and in fact their wheels often leave the ground for some fractions of a second.

Just beyond the hill lies the entrance to Corners 8 and 9, also called The Esses. Trying to set up for a turn just as your vehicle begins to take off is not a comfortable feeling. Motorbikes have no ailerons; they can't be steered in the air. Yet coming over the top at Corner 8 you may be in low-altitude flight. For a moment you have about as much control as a bug sitting on a pebble that someone has flung into the air.

Corners 2 and 5 are the "glamorous" turns at Mosport. People may also talk about The Esses or The Chute, and they usually have a few anecdotes about Corners 1 and 10.

The first corner—at Mosport or at any other racetrack—is always a zoo at the start of a race. The last one (Corner 10 or "White's Corner" at Mosport) is everyone's final chance for passing or being passed before the chequered flag. Most racers have "reeled in" someone, or have had someone "motor by" them, at the exit of Corner 10 on the last lap.

But Corner 3 is not glamorous. Few people talk about Corner 3. Although this fast, sharp, complex right-hander is just thirty-nine degrees short of a hairpin, it doesn't really scare anyone. It may be because everybody finds it such a relief after Corner 2.

Corner 3 can be strategically important in a close race. It offers good opportunities to pick up a position at its entrance or its exit, with little danger of being re-passed for the next two turns. A racer who takes the lead out of Turn 3 can often stay in front through The Chute and into Moss' Corner where passing can be difficult. If he gets a good drive out of Moss' Corner, and his machine has horsepower, competitors may find it hard to catch him on the back straight. On the final lap this means that he has only The Esses and Corner 10 to worry about before the chequered flag.

Corner 3 is actually two corners, A and B, with a radius change at roughly 500 feet into the turn. At this point the radius lengthens from 300 to about 570 feet. The angle of the turn also becomes milder, shifting from ninety-five to about forty-two degrees. Turns of this type are sometimes called "increasing radius" curves and are generally regarded as easier to go around than "decreasing radius" curves (curves whose radius shrinks so that they become progressively tighter). In fact, as physical entities, both types may share the difficulty of having two apexes. Otherwise one is neither easier nor harder than the other. They simply require different racing lines and techniques.

However, as mental images, shrinking-radius curves are spooky in a way that stretching-radius curves are not. That's another reason for Corner 3 being less glamorous. If you stuff it in Corner 3 you tend to shut up about it. Losing it in Corner 3 doesn't make you look very good. In fact, it may make you look like something of a fool.

On Saturday, September 23, 1967, Frank was planning to cut a few laps at Mosport. The track fee back then was about $20 for a day's "open test", even for those racers who weren't fellow Czechs or acquaintances of Jerry Polívka, then part-owner of the circuit. Wally Esden had put together a really fast Yamaha 250 for a customer and agreed to have Frank set it up at a racetrack before delivery.

Frank was looking forward to the test. The new screamer

was exactly like his own Yamaha from the season before, only potentially faster: a chance for a racer in search of the Holy Edge to pick up an idea or two.

It was a fairly pleasant day, but late in the season. There weren't many riders taking advantage of the open test. John Nelson turned up and so did Tom Faulds—still a young scratcher in those days and not yet vice-president of Honda Canada. Jerry Polivka came in his brand-new car, a Mustang with white leather seats. One sidecar guy showed up with his rig and his passenger. While sidecars usually run separately from solo riders, there was no point in splitting the practice sessions this Sunday. There would be no more than eight or nine people at the same time on the course.

Frank's crew consisted of his German Shepherd and a young man who worked in the parts department at the shop. Just a few days earlier the parts department was stocking up on the new full-face Bell helmets, available for the first time in Toronto. Frank liked the look—real space-age stuff, even if a little heavy—but there was no way he would spend $80 for a helmet. It was very nearly his rent for a month. Now he put on his own Bell, an old open-face model, with a tight leather mask that could be pulled up to cover the chin and the mouth. It wasn't much protection but it kept out the wind and the bugs.

He slid a little elephant-figurine inside his helmet, as he always did. It had been a gift from Jana for good luck—not that Frank was in any way superstitious.

Few racers—except maybe British world champion Barry Sheene—will openly admit to being superstitious. After all, a racer's trade is close to engineering, a true applied science, with side-trips into chemistry, metallurgy and physics. So Frank was no more superstitious than the rest of his fellow scientists. He only liked to keep away from the number thirteen, and maybe from racing in brand-new leathers, boots, or gloves. And well, yes, he'd always stick Jana's elephant inside his helmet.

The Yamaha started easily. Frank kept a cautious finger on the clutch, like everybody who races a two-stroke. Highly stressed two-strokes have a nasty habit of seizing, locking the rear wheel, and depositing the unwary rider on his ass. But this time the little Yamaha's engine spun freely. It also pulled like a locomotive once it came on the pipe. Keeping it in the revs was a breeze. Frank never liked to go slowly and on this bike he didn't have to.

A few drops of rain fell—Mosport has its built-in cloud, parked right over the circuit even when the rest of the sky is clear—so people decided to break for an early lunch. Shortly after noon the rain stopped, the sun flickered, and the track dried quickly. Frank started up the Yamaha again. He wanted to test the suspension.

After a lap or two he was getting into the rhythm. Deep into Corner 1, then hug the right. Drift left across the track, then aim right before the hill. Hugging the right, peel off left just before the crest, then let it drift deep into Corner 2. Keep it cranked over all the way and it will set itself up on the left for Corner 3. Now peel off right, keep it cranked over and . . .

Suddenly the exhaust pipe on the right side touched down. The rear wheel became light. Frank wasn't too concerned at first; the pipes had touched down before, and the rear end can tolerate a little air under the wheel. A small drift or hop can even help point the the bike in the direction you want it to go. Then you can let it stand up a bit, reduce the lean angle, pull back a tad from the outer limits of traction. He was explaining the point to someone just the other day in English.

"You know, when rear wheel lifting," he had said, "you not counting that front wheel go up too."

But on the afternoon of September 23, 1967, the front wheel came up. The Yamaha began sliding on the pipes. It was sliding toward the pile of rocks on the left side of the track. That was the last thing that Frank could remember.

The sidecar guys were running about half a lap behind Frank, so they didn't get to Corner 3 until a minute or so after the crash. They slowed, looked, but didn't even pull off the track. When they got back to the pits and reported what they had seen, someone immediately hauled out the red flag.

Practice stopped. When all the riders were off the track, John Nelson and Tom Faulds piled into Jerry Polivka's Mustang. On open test days there was no ambulance standing by in the paddock.

In Corner 3 everything was quiet. The Yamaha had evidently slid across the track, scrubbing off speed as it went. When it slammed into the rock pile it was probably moving at no more than 50 mph. It had been sliding on the lowside, and Frank appeared to have stayed with it all the way. When the bike hit, the impact flung him through the fairing face-first into a large rock.

The $80-Bell helmet would have helped. The Fiberglas chin-guard might have absorbed at least part of the impact. As it was, Frank had no nose left. A piece of plastic windscreen was sticking out of his face. Mercifully, he was unconscious.

It also seemed that Frank had broken a number of bones, certainly in his ankle, possibly in his neck or back. John, Tom, and Jerry briefly debated the wisdom of moving him, but there was no alternative. Some fifteen or twenty minutes had elapsed since the mishap and Frank was losing blood. No ambulance could reach the track in under forty minutes. By then Jerry's convertible could have Frank in the hospital at Bowmanville.

The Mustang's white leather seats were red as they pulled up at the emergency entrance. The orderlies lifted Frank onto a stretcher and rolled him into a corridor. John stayed with him, while Tom and Jerry drove back to Mosport. For the moment there was nothing more they could do.

Some nurses and a young intern looked at Frank briefly, but then all activity ceased. After nearly two hours of waiting John Nelson started freaking out. A resident doctor came to see what the fuss was about. He was sympathetic, and eventually helped to arrange for an ambulance transfer for the patient to St. Michael's Hospital in Toronto, but he appeared to be in agreement with the rest of the hospital staff.

The doctors were busy with people they could help. This patient, apart from all other injuries, had obviously broken his neck. By then it was swollen to a thickness greater than his head. A purple tree-trunk jutting out of a pair of shoulders with next to no face. He was still breathing quite well, considering, but he was not expected to live.

13

GOING IN DEEP

Jana had said her good-byes to everyone at the Ice Capades in Los Angeles. She bit back her tears, made all necessary mental adjustments, and travelled to North Bay to start her new coaching job. It would probably be safe to say that she was not in the happiest frame of mind, but Jana had always been a trouper. Having made her choice, she was prepared to make the best of it.

For a start, she spent the day finding herself a flat in the bleak little town in northern Ontario. The plan was for Jana to stay in North Bay on her own, with weekend visits from Frank, until a coaching or modelling job came up in Toronto.

It was Saturday, September 23.

Frank, as she knew, would be spending the day at Mosport, testing a bike for a Checkered Flag customer. Jana did not expect to speak with him at all until she called him with her new phone number. Since nobody had Jana's address in North Bay yet, she was quite surprised to hear her phone ring around 11 o'clock at night.

1967. Frank poses at Mosport astride the Yamaha TD-1C before his crash in Corner 3. Note the low pipes.

1968. Motocross as physical therapy. Frank on his way to winning the Canadian Open Championship (Junior)

1967. Ice Capades, Los Angeles

1972. Crazy Frank wiggles on the H2. Note the hammered exhaust pipes.

1974. Moss' Corner. Chris Manley (11) chases Frank's Green Monster.

1975. Front-row start at Mosport. Polesitter Number 1 is Steve Baker, glancing at Mrazek. Did you jump that flag, Frank?

rvell
BMW
Bavarian Motor Works
SILVERSTAR
OTTAWA CANADA
STP
THE RACER'S EDGE
ski-doo
gaf
FILM
40
gaf

1976. Frank in the paddock at Mosport after beating Steve Baker in the Canadian Grand Prix. Richard Chambers is on the left.

1977. Frank on his victory lap at St. Jovite aboard the OW31

1980. Mosport, Corner 2. Frank is piloting the Honda's monstrous six-cylinder CBX to victory in the Open Class.

1988. High priests of speed. Frank (77) on the Classic Racers Canada Ducati and Paul Bowyer (40) wait for the flag at Shannonville.

1988. Frank's team, Classic Racers Canada, after Frank has won the Canadian BOTT Pro Open title in the Eastern Canada Challenge. From left to right: Terry Wolfe; George Jonas; Ducati tuner Tim Spiegelberg; crew members Tom Glasser and Maya Cho; Triumph tuner Larry Rose; Frank Mrazek. The mutt's name is Charlie.

It was the police. They told Jana that her husband had been injured in some accident but could offer no other details.

The police found Jana because Frank's crew at Mosport, the young fellow from the parts department, had been smart. He only knew Frank as a mechanic at the shop, but somehow he remembered that Frank was friendly with the Jelineks. He loaded the German Shepherd and the wrecked Yamaha into the van, drove back to Toronto, then called Otto Jelinek. Jelinek knew that Jana was supposed to be somewhere in North Bay and he phoned the Ontario Provincial Police.

Jana jumped into her car. Later she couldn't remember much about the trip, but by around 1:30 a.m. she was at Toronto's St. Michael's Hospital talking to the doctors. Considering the distance—some 200 miles—she must have been flying.

For the next eight days no one could tell Jana much because Frank remained unconscious. At that point the doctors weren't worried about the two broken legs and the shattered ankle. They also felt sure that they could reconstruct the nose and fix the facial bones. As for the back and neck—well, as it turned out, only some of the transverse processes, the bony projections that jut out from each vertebra, appeared to be fractured, which was not crippling or life-threatening. However, possible trauma to the brain was something else. The medical people in Toronto were not much more optimistic than their colleagues in Bowmanville had been, albeit for different reasons. Frank would have to regain consciousness before the doctors could offer Jana a prognosis.

On the ninth day Frank came to, which he signified by asking the nurse, "You know if Yamaha is bang up pretty bad?" The nurse thought her patient was raving. Frank couldn't remember anything about the accident, but he recognized that he was in a hospital and he could put two and two together. He had woken up in hospitals following motorcycle events before.

Still, on previous occasions he had never had to stay in

hospital for eleven months after waking up. This time he did. Just to fix the nose required several operations. First with a wire, and when the wire started coming through the skin with a piece of plastic, and when that didn't work either with a chip of his bone taken from the thigh. The broken legs and the broken cheekbones healed by themselves, though Frank was never happy with how his face looked afterwards. He used to have a "skinny nose" as he pointed out to everyone later, and he regretted not buying the full-face Bell for $80. "If I have that helmet, I be still good-looking guy," he told the nurse.

But the real problem turned out to be the ankle. The joint had simply shattered on impact. Even though they fixed it, first with steel pins and eventually with a series of artificial joints made of plastic, with four operations in all over the next number of years, it never became fully functional. The ankle would leave Frank with a pair of crutches for about fifteen months and a heavy limp for a lifetime. Worse than the limp, it would leave him virtually unable to operate the left foot-controls on a motorbike, which meant the rear brake on some British and Italian models, and the gearshift on the rest.

Walking even with a limp was still a miracle because the doctors at first didn't think that he would walk at all. Frank was actually very lucky with his doctors. As it turned out, Dr. Eduard Kosinka from Brno, the very guy who used to fix up Frank after his spills in Czechoslovakia, had also made the trip to Toronto in the meantime. Motorcycle racers should always import their doctors when they change countries, and even though Dr. Kosinka wasn't exactly imported by Frank, it amounted to the same thing.

Then, there was his new Canadian medicine man, Dr. John Evans. Between the two of them they managed to wire Frank together, almost as good as new. True, Frank helped, partly by being stubborn and not letting anything stop him for long, and partly by being a pretty quick healer with a high pain threshold. A whopping contusion or a deep cut that might scar other people for a long time would disappear from Frank's skin within a few weeks.

Dr. Evans was miraculous, a real lifesaver. Miraculous, not only in that he was good at his job, but also in realizing that his job wasn't to change patients but to fix them. He could rebuild Frank when necessary, but he didn't worry about making things easier for himself. He didn't fret about Frank following his orders, or doing only what was medically wise. Unlike many doctors who get off on telling you how to live your life, Dr. Evans and Dr. Kosinka saw their challenge in making it possible for Frank to live his life the way he wanted to live it.

The way Frank wanted to live his life was, of course, by racing. Between races—well, working on motorbikes, then eating, drinking, partying, which went with the territory. These were the things that Frank wanted, and what he was willing to give in exchange for them was everything else. Certainly minor things like money, comfort, security, and life.

The only thing he would have preferred not to have to give in exchange was his marriage. Or his hopes for a family. It was normal for a guy to have a wife, and later maybe one or two children. Actually not just "a wife" but someone he happened to be in love with—in Frank's case, obviously Jana. That had not changed since the day he first saw her picture on the cover of the *Stadion*. The problem was, Jana had become totally opposed to Frank's racing.

As far as Jana was concerned, the only good thing about the accident at Mosport was that it would stop Frank from setting foot on a road-racing circuit again. Partly because he couldn't—the man could hardly *walk*—and partly because no one would give him a ride. During the eleven months that Frank spent in the hospital The Checkered Flag had gone out of business. Dragging himself on crutches Frank would be lucky to get any kind of a job again, let alone a new sponsor.

Jana was never interested in racing. She'd go to the track once in a while to time Frank's laps or wipe his goggles because, well, it was what women did when they were married to racers. But while she wasn't interested, she used not to be worried. She knew that racing was dangerous, naturally, but it was a remote kind of knowledge. It didn't spill over into her feelings at all. Perhaps it was because Frank had generally been lucky. If now and again he'd hobble for a few weeks or wear his arm in a sling, it didn't seem to matter. In Jana's circles a lot of people had sports injuries.

But this thing at Mosport was different. This crash left Frank a cripple and nearly cost him his life. Perhaps racing was just more dangerous in America. Perhaps the speeds were higher or something, but it certainly wasn't a sport that a responsible married man should do. And (though Jana didn't say this either to Frank or even to herself at the time) a man shouldn't do it especially after he had made his wife choose between *her* sport and marriage.

In any event, when Frank started to talk about going racing in the late summer of 1968, Jana thought that he was joking. She actually laughed because Frank happened to raise the subject as she was handing him his crutches, and she thought that it was funny. She never took him seriously for a moment.

Perhaps Frank himself wasn't serious about it at first, though by then he was going to motorcycle races again. They were races of a different kind, though, held in farmers' fields in the dirt.

Frank never thought that he'd be interested in motocross. Like most youngsters he started out by riding cross-country enduros, but they did not excite him much and he felt that motocross would be about the same. Motocross wasn't exactly cross-country, but it involved going around on a short track laid out in a field that utilized the contours of the landscape. Sometimes the track was enhanced by a few artificial hills or mounds for jumps. The speeds were relatively low, and the riding style was totally different from road racing. People would stand on the footpegs for the rough parts and for the jumps, and on the corners they would sit up straight, stick out a foot to anchor themselves, then power the sliding rear wheel around the turns.

Motocrossers were exciting to watch as they sent up huge rooster tails of dirt going around the corners. The dirt would soon form a kind of earthen curb around the outside of the turns, propping up the rear wheel—"shooting the berms", as the technique was called. Frank picked up on all this only because of his new job, which wasn't really a job but a business: selling parts, tires, and accessories for CZ motorcycles from his house.

The Czech factory was building scramblers for motocross racing, and it had established an outlet in Montreal called MOTOKOV. Racing in the dirt called for the development of special machines. They had gradually evolved from all-purpose scramblers to very tall, very light contraptions that started to look a bit like plastic giraffes. They had knobby tires, metal skid plates, wide handlebars for leverage, high pipes, and footpegs for ground clearance, and stood on forks that seemed to go on forever. It was necessary to have such long suspensions to absorb the tremendous forces upon landing after a jump.

Anyway, Frank knew some people at MOTOKOV and they gave him a few tires, spare parts and a couple of complete CZs, a 250 and a 360, so he could hitch a trailer to his car and sell parts to motocrossers at the races. He was still on crutches, but being a kind of travelling salesman was a way to pick up a few extra bucks. Luckily by then Jana had a good job coaching in Toronto.

The dirt guys kept teasing Frank that he should go with them for a practice session or two, it would soon put him in shape again. Everybody laughed because it was a joke. Motocross is probably the most demanding sport on earth. There is nothing to touch it in physical terms; the only thing that would come close is the decathlon. On medical tests for all-around conditioning it is generally motocrossers who finish ahead of everybody else. This is not surprising: to duplicate the effort required for motocross one would have to keep jumping off three-foot walls non-stop for about fifteen minutes, holding a ten-pound weight in each extended hand.

One day Mike Manley, a road racer just like Frank, happened to come by Lloyd's farm in Guelph, which was one of the local tracks. He also happened to have a 450 Honda scrambler in his pickup. "Hey, Frank," said Manley, who was then a Honda distributor, "maybe you and I should try this."

"No," Frank shook his head. "I never sit on this thing before."

"Well, neither have I," said Manley. "So what? Are you scared?"

This was pushing Frank's button. "I have no bike, so what I race?" he asked.

Manley pointed to the big 360 CZ on Frank's trailer. It was brand new, a sample bike MOTOKOV gave to Frank, certainly not with the intention of putting it on the track at Lloyd's farm for fun. However, as Manley started to unload his Honda, this became a secondary consideration. Frank cut four oval plates out of cardboard, then used some old motoroil to paint racing numbers on it for Manley and himself.

They started out with the juniors—it would have been insane to go in the senior class, to say nothing of the experts. It was crazy enough in any event. But motorcycles were motorcycles, and right after the start Frank found himself dicing with Manley in front. Neither of them knew what they were doing, but it happened to be an easy track. It had a lot of fast turns and only a couple of jumps, so even though Frank hated putting his foot down in the corners and kept landing on his front wheel coming off the hills—he had never tried jumping before—he somehow won the race. Manley came second, so the two of them beat all the young motocross guys the first time out.

Manley wasn't too happy with second place, though, just as Frank wouldn't have been. They were both winners by nature. "I musta been stupid, talking you into this race," Manley kept grumbling as he put his Honda back into the van.

Frank also loaded up the CZ and drove home. It was only when he was unloading the trailer at his house that he noticed he hadn't been using his crutches all day. That was good. It was very good, but he still considered it more prudent to say nothing about the race to Jana.

It wasn't really a decision, because in Frank's life things sort of decided themselves. Maybe they do in many people's lives—more often, at any rate, than many people would like to believe. The way Frank side-stepped the problem as he started racing motorcycles again was by saying to himself that he wasn't actually *racing*, he was only doing motocross. This became his position to Jana as well. Curiously enough, she accepted it at first.

Frank wasn't racing, you see; he was just doing what he had to as a MOTOKOV sales and service representative. A guy has to make a living. Also: Frank wasn't racing, because racing was going fast on hard asphalt rather than riding slowly in soft dirt. What he was doing was, well, physical therapy.

It was interesting. Jana could maintain the illusion that motocross meant riding "slowly" in soft dirt only as long as she didn't see any of the races—and since she didn't go to see them, this was fine. But the "physical therapy" part was something else. It was both completely true and absolutely ridiculous.

It was true because the effort involved in motocross had soon turned Frank from a rather slender guy into a beefy hulk. Hard as rock, too, bulging with leg and upper body muscles. At the same time it was ridiculous, because for someone to say, "Well, I'm in no shape to do road racing so I'll do motocross," was like saying, "Well, I'm in no shape to jog around the block so I'll swim across the Dead Sea."

At thirty-two Frank was past the age where people could begin doing motocross at all, with or without a crippled ankle. Not that road racing requires no physical conditioning—contrary to what many people think it is a very strenuous sport—but it still isn't anything like motocross. The normal progression for competitors is to do motocross from, say, fourteen to twenty-five, then go into road racing. And even if they continue doing motocross into their forties, as some do, they don't start it at thirty-two. For a former road racer to go into motocross at that age—go into it on a serious, competitive basis—is pretty much unheard of.

Well, to make a long story short, Frank did it. He became Canadian Open Class Motocross Champion (Junior) in 1968 and then again in 1969. Whether or not it was good physical therapy was debatable—he kept shattering the metal pins in his ankle until Dr. Evans replaced them with the first of a series of plastic joints—but it was excellent mental therapy. It eased the transition back into racing for Frank and even for Jana; though she soon discovered that while people may not wipe out in the dirt quite as hard as they do on the pavement, they wipe out more often. There was hardly a weekend without Frank hobbling through the door with some new cracks, cuts, and bruises.

Cuts and bruises couldn't have mattered less to Frank. What mattered to him was that he still had it, that he hadn't lost it at Mosport in Corner 3, "it" being something far more important that a nose or an ankle. It was—well, Frank didn't particularly cherish spelling out what "it" was any more than other racers would have, but you could probably call it "guts" or "heart" if you liked. You could say it was the willingness to go for it, to seek that feeling, that high, that mystery.

"It" was, for instance, when Frank discovered in Welland how to make one jump out of three hills on a certain whoop-de-do. He decided not to go jump-jump-jump like all the others but jumped directly from the first hill to the *third*, flying over a second lower hill, and the heads of everybody else, on the track. This was when spectators rushed to the spot, saying, in Frank's translation: "Come on, let's look it, Frank go get killed there," as he was leading the race by half a lap. What "it" was, then, was doing whatever it took to win.

On a more practical level, motocross had made Frank a factory racer, at least by proxy, for the first time in his life. MOTOKOV gave him the CZ 360 to race, along with a supply of parts and equipment. They paid for his racing efforts and even distributed a calendar with his picture on the cover.

In North America Frank became what he could never become in Czechoslovakia: a works rider for CZ. Suddenly it didn't seem to matter anymore that his father had been a kulak. By defecting he had achieved what he could never achieve by staying put like a good soldier. Hell, his old country might even have been willing to make him a second lieutenant, now that he was 3,000 miles away from home.

While Frank had been busy defecting, cooking chocolates, learning English, and then recovering from his injuries, a couple of revolutions had occurred in the world of road racing. One had to do with riders, the other with machines.

Perhaps "revolution" was the wrong word because they were really developments, occurring over a period of many years. Still, they seemed revolutionary to anyone returning to the track, as Frank did eventually in 1970.

First, in terms of machinery, road racing had all but become Japanese. That had not happened overnight. In the 1950s when the Oriental factories first brought their products to the Grand Prix circuits, mainly in the tiddler classes, Europeans laughed. It took them until the early 1960s to stop laughing, and even then they tended to keep superior smiles on their faces.

In the early sixties it still looked like Europe as before, Europe as ever. Count Corrada Agusta was doing what Count Giuseppe Gilera had been doing between the 1930s and 1950s, or, for that matter, Adalberto Garelli had done during the 1910s and 1920s. To say nothing of Moto Guzzi, Ducati, Benelli, Norton, Triumph, Matchless, JAP, Velocette, Vincent, AJS, Ariel, Rudge, Royal Enfield, Excelsior, Sunbeam, BSA, NSU, BMW, DKW, Puch, Morbidelli, Aermacchi, Jawa, CZ, MZ, and many other dazzling marques throughout the European century of this essentially European sport. With the exception of some respectable competition by America's Harley-Davidson and, earlier, Indian racing bikes, the (mainly) British-Italian-German rulers of the Grand Prix circus did not feel threatened by anyone.

By 1970 all smiles had faded. Europe was trying to hang on, grimly and not very successfully. Japan's miraculous rice-burners were simply running away with the sport. The bikes wheeled down Victory Lane now tended to be Suzukis, Yamahas, Hondas, and Kawasakis, probably in that order: not yet exclusively, but predominantly. Japan was beating Europe at its own game.

The top riders, on the whole, were still European along with several Australians, New Zealanders, and Rhodesians. It was New Zealand's Hugh Anderson and Germany's Ernst Degner who helped to put Suzuki on the racing map. Riders who did the same for Yamaha included Britain's Bill Ivy, Phil Read, and Rod Gould, as well as Finland's Jarno Saarinen. Australia's Tom Phillis died trying to do it for Honda in 1961. Later it was Rhodesia's Jim Redman, Italy's Luigi Taveri, and, of course, Britain's "Mike The Bike" Hailwood who gave Honda some of its greatest victories.

North Americans were not yet represented on the international road racing scene in significant numbers, though Canadian works riders Mike Duff and later Yvon DuHamel did well for their Yamaha and Kawasaki factory teams. Hard as this is to believe today, U.S. riders were not then thought of as top road racers. By the end of the sixties the Japanese factories had the winning machines, and could pick and choose from among the best drivers in the world.

The question of the best drivers involved another revolution, a revolution in racing style.

Some people later said that it all started in Australia. Others maintained that it was essentially a personal discovery by British ace Paul Smart. Still others pointed to racing photographs of John "Moon-eyes" Cooper dating from as early as the mid-1960s, along with pictures of Hugh Anderson, Frank Perris, Giacomo Agostini, Cal Rayburn, Jarno Saarinen, and others, all showing an unmistakable knee sticking out to the inside of a turn.

The new style of "hanging off" probably did not evolve from an individual's invention, but from a collective search by several top riders looking for a winning edge.

The classic style demanded that a rider sit straight "up-and-down" as though he were cast of a single piece with his machine. He would lean into a turn still holding his body in a line parallel with his bike while gripping the tank with both knees. Naturally, he would lean as far as his nerves would allow, since the further he leaned his bike the greater the speed at which he might go around a corner.

However, there were limits to lean angle regardless of a rider's daring since no one can lean a bike further than it will go. At one point all bikes run out of ground clearance or traction, or sometimes both together. Championship drivers with equal nerves and machines had to look for an edge elsewhere, and some of them instinctively discovered it in weight transfer.

By sticking out a knee and hanging off the motorcycle, a racer could slightly reduce his lean angle and still maintain the same cornering speed. By lowering his body beside the bike he could shift the centre of gravity to the inside of a turn and diminish the centrifugal forces acting on his machine. By dragging a knee along the tarmac he could increase his speed and stability and extend the limits of traction and ground clearance over his competitors'.

Though the style looked spectacular, the edge offered by hanging off was not as dramatic as it appeared. Superior racers with faster machines could still run away from the pack without hanging off, but among riders of equal ability and equipment the new style favoured the knee-draggers by a few tenths of a second, and that was often sufficient for victory. By the late seventies nobody could hope to win a Grand Prix in the classic style—well, almost nobody, because in 1978 Mike Hailwood still won the Isle of Man TT sitting straight up-and-down. However, Hailwood was probably the last racer to do so. Within a year people could hardly win at club level without touching their knees to the pavement.

The new style looked, and was, somewhat acrobatic. It required a degree of physical fitness. To drastically shift one's body from side to side on a bike, whether at speed or under heroic braking—and to shift it without any unwanted steering input—required not only daring and experience but a fair bit of abdominal muscle. It also required a mental adjustment many classic-style riders were not prepared to make.

Luckily for Frank, he was not a theoretical racer. He wasn't wedded to any style. He rode with his instincts and his sinews, always ready to try something new. When Kawasaki offered him a ride in a twenty-four-hour endurance race at Harewood, Ontario, he looked at the two or three early birds who were already hanging off, said, "Hey, I like that," and just started doing it.

Hanging off worked like a charm for him. There was nothing to it. It seemed to make Frank go a little faster, which was the name of the game. He was in good shape from motocross and didn't need to use his arms for leverage when shifting his body, which was one of the things that discouraged some racers from hanging off because even a slight pressure on the handlebars could jolt the bike out of line. Frank enjoyed the feeling of scraping the pavement with his knees. Nobody had knee-pucks then, so young scratchers just put some duct tape on their leathers.

In the early seventies knee-draggers were still a minority. Even those who started hanging off did not yet do it in an exaggerated fashion. That came later. By the 1980s many racers would slide off their seats until their hips were roughly in line with the opposite footpeg. They'd virtually squat or kneel beside their bikes on the inside of a turn, but this extreme style did not develop until the appearance of wide racing slicks—huge, "bald", soft-compound tires for maximum dry grip—in the middle of the 1970s.

Revolutions often work for outsiders, and the racing world's revolution worked for Frank. First, by adapting to the new style ahead of other riders he could be competitive. Second, because of the motorcycle market's response to the Japanese invasion, he could start getting rides.

Not *factory* rides, though. Works rides became, if anything, harder to get once the Big Four of the Orient had the pick of the field. They didn't have to look for little-known Czech veterans of thirty-four. But a consumer revolution followed the

Japanese victories at the Grand Prix circuits, and it made competing distributors, dealers, and after-market manufacturers eager to grab a slice of the exploding home markets. Many of them began sponsoring racers and racing teams, primarily in the production classes, to attract new buyers to their brands.

This signalled the start of a golden age for top privateers. For Frank it began with Kawasaki's fearsome smoking triples.

Towards the end of 1969 Frank had started working as a mechanic for Mike Manley who by then had shifted from Honda to Kawasaki Canada. Motocross had been loads of fun, but in truth the dirt was, well, *dirty*. It was perfect for growing cabbages, but for racing bikes the asphalt seemed better. Frank had always been a bit of a neat-freak and he liked to have clean machines. It bothered him that a chimneysweep looked like a bridesmaid compared to a motocrosser after an hour in the field. So, when Manley asked Frank to ride a new H1 for his Kawasaki distributorship, he did not have to ask him twice.

The legendary two-stroke three-cylinder Kawasakis, designated H1 and H2, displaced 500 and 750 cc's respectively. Later in racing trim they were called H1R and H2R. In terms of power they were awe-inspiring from the first year, and very popular with younger buyers. They were also awe-inspiring in terms of handling, but in a different way.

"That thing was *wiggle*, and nobody can stay on it" was the way Frank would remember the vehicle of his first road-racing triumph in the New World. "My bike never any handle, but I have the nerves to keep wiggling."

Frank wiggled his way right into becoming Canadian Champion in the Open Production Class in 1970 and in 1971. In the 500 cc G(rand) P(rix) Class he finished in second place. He also began competing in endurance races: on the winning Kawasaki teams in the twenty-four-hour and six-hour events, and in a twenty-four-hour race on the second-place finishing BMW team.

On his Kawasaki triples Frank managed to wiggle his way

into a new nickname as well. People began referring to him at North America's racing circuits as "Crazy Frank".

Just as there are many different sports that ritualize personal combat—boxing, wrestling, fencing, judo, and a number of others—speed contests also come in a variety of ways. They can be as different as fencing is from wrestling. Road racing, motocross, drag racing, trials, enduro, dirt track racing, hill-climbs, ice racing, and the rest require very divergent skills and techniques.

Even within contests of the same type there are major differences. Road sprints of ten to twenty-five miles, held on tracks from one to four miles in length, are hardly the same as Tourist Trophies (TTs) laid out on courses ranging from ten- to forty-mile laps. The first can mean racing in a traffic jam; the second, racing against the clock with close competitors never even catching a glimpse of one another.

There are special events, such as the famous Daytona 200, consisting of fifty-seven very fast laps through a tricky infield and two 31-degree banked ovals. Grands Prix, depending on class, go from just under forty-five to just over ninety miles. They are usually held on international glamour-circuits of about two to four miles in length, such as Paul Ricard and Le Mans (France), Assen (Holland), Imatra (Finland), Anderstorp (Sweden), Silverstone (Britain), Jarama (Spain), or Imola, Mugello or Monza (Italy), and others. Grands Prix tend to run about forty-five minutes.

In contrast, endurance races can last anywhere from two to twenty-four hours. They're generally between teams of two to four riders, and may require a competitor to drive at racing speeds for up to three hours at a time. (Sane people wouldn't wish something like that on their worst enemies, but many racers clamour for the privilege.)

Contests are split into various classes of riders and machines. Riders will usually compete as amateurs or professionals, novices or experts, juniors or seniors. Whatever their ranking, they must be licensed by a national sanctioning body, affiliated with, or at least recognized by, FIM.

The bikes are classified by engine size and type, somewhat like boxers are by weight-groups. The "heavyweights" may be 750 cc to unlimited or open class bikes, the "middleweights" 400 to 750, the "lightweights" up to 400 cc's. Their classification may also be affected by the number of cylinders, engine cycles, cooling methods, and so forth. In addition, bikes in the Production/Sport classes must be "street" motorcycles. They are homologated according to certain specifics, usually the number of units available at commercial dealers.

The system is highly complex. The rules may change from season to season as technology advances, but in each period a premier or glamour class emerges, just as in professional boxing the headliners are often heavy- or middleweights. In Grands Prix, tradition still gives the premier position to the 500 cc class. (Other traditional GP classes are 350, 250, 125, 80, and 50 cc, though some have been dropped while at one time a Formula 750 class had been added.)

What GP and Formula classes have in common is purebred racing machines: often factory entries, and quite distinct from motorcycles one can buy at the local dealer's. At the other end of the spectrum, in the Production/Sport classes, bikes have to stay virtually unmodified. Riders can remove lights and mirrors and maybe add a steering damper, but that's about all.

Production/Sport classes are always popular, while GP and Formula classes are magnificent. Twin-cylinder motorcycles, light- or heavyweight, form an elegant class by themselves. However, beginning in the early 1970s, and especially in North America, the glamour has increasingly shifted to Superbikes.

Superbikes started out as production-based machines of 750 cc to unlimited displacement, often heavily modified for racing. Eventually their range extended down to middle- and even lightweight "superbikes" of only 400 cc's. Today they still don't have the patina of GP and Formula racers, nor are they part of the official FIM world championship classification, but they're probably the most hotly contested and popular class. They have all but replaced Formula One and 500 GP as the premier class in national events in North America.

The appeal of Superbikes is based on their awesome performance, probably combined with their apparent similarity to bikes any spectator can have in his or her garage.

In 1972 the International Grand Prix circus was pretty much a closed shop. It was a glittering circle that aging privateers like Frank had little hope of breaking into.

However, Superbike racing was the new kid on the block. With Superbikes Frank could get in on the ground floor. When Kawasaki Canada offered to let him ride its new Z-1 1200 cc open-class Superbike, nicknamed the "Green Monster", Frank said yes before they could finish the sentence.

The Green Monster was not easy to tame. It could come up with some nasty wiggles of its own. Later that year it reminded Frank who was boss by throwing him for two broken legs, eight broken ribs and two broken collarbones at Nelson Ledges, Ohio.

It was not as bad as Mosport, but it came close. Frank had some doubts about the state of the medical arts in Warren, Ohio, the nearest town to the track, so he had himself smuggled back to Dr. Evans on a stretcher in his teammate Derek Mitchell's van. It wasn't even an unpleasant trip, considering. The Ohio doctors must have pumped the entire morphine supply of the state into Frank before squeezing his stretcher between a couple of bikes, some gas cans and a lot of spare parts.

Luckily it was late in the season and Frank could take half the winter to heal. Officially he even "retired", telling Jana that this was it, period—except he didn't believe it and neither did she. Things were going too well. For instance, Mike Manley had lent him three Kawasaki 350s so he could set himself up in business as a dealer. True, other than three borrowed bikes, Frank had only $17 in the world at the time, but he had his tools, could rent some space in the Toronto suburb of Mississauga, and soon there were customers showing up for modifications and repairs.

People had heard about "Crazy Frank" and they would bring him their bikes to fix. To fix and, preferably, to make a bit faster. If they showed up during the week they were not disappointed, but soon Saturday customers were greeted by a sign that said: GONE TO RACING, SEE YOU AT MOSPORT.

What was a guy to do? How could he pass up a chance to team up with Yvon DuHamel and Ted Redford and ride a Kawasaki 900 Z-1 to victory in the twenty-four-hour endurance race at Mosport? And, since he was racing anyway, how could he forgo winning both the Canadian (CMA) and the U.S. Western/Eastern Racing Association (WERA) championships in the Unlimited Grand Prix and the Unlimited Production classes for 1973?

In 1974 Frank repeated the performance. He was now twice U.S. and Canadian champion in two classes. Teamed up with Chris Manley, Mike Manley's younger brother, he also won at Mosport the 1000-kilometre endurance event on the Green Monster. By then Frank was racing for the Lester Team as well, owned by Tom Lester, the Akron, Ohio, manufacturer of America's popular after-market wheels, and a fine gentleman to boot. The Lester Team won the five-hour and twenty-four-hour endurance events at Nelson Ledges, so next year Lester made Frank team captain. After all, at thirty-eight he was the oldest guy on the team.

If Frank had occasional doubts about the way his life was going, it had less to do with himself than with Jana. For his own part, sure, he might get tired of travelling or never having any money, but it was still a small price for that *feeling*. The feeling that nothing else in the world—nothing—could replace.

For instance, Mike Duff had retired by then. He was running a Yamaha dealership near Toronto, and Frank would listen with his eyes wide open in astonishment as Duff talked about his new hobby, which Frank understood to be nature photography.

Duff had gone back into racing as a privateer after his accident in Japan. He still beat Bill Ivy at Mallory Park's Race of the Year in England and came second to Giacomo Agostini at Brands Hatch. This was world-class racing, so Duff hadn't been slowed down by his crash. Yet, when he couldn't get another factory ride, eventually he retired. Retired for good, which puzzled Frank to no end.

Perhaps, never having been so near the top, Frank just couldn't understand how difficult it must feel for someone to be on the way down. Or perhaps to him it wouldn't have mattered, up, down, or sideways, as long as it was on the racetrack.

"You selling bikes. You here between motorcycles, you must go crazy," Frank remembered saying to Duff. "How can you live?"

Duff just shrugged. "Frank," he said, "I don't give a damn about it."

Frank was amazed. This was a guy who could beat Bill Ivy and chase Agostini. Imagine, Agostini! "Go to somewhere in forest and take picture from some *birds*, I'd be going nuts," Frank would say later. "From that exciting stuff, from that exciting racing, go there? It's like you're dead."

Some people, of course, were dead anyway by then. Killed racing like the CZ factory riders Jirka Kostyr and Standa Malina, or killed just riding in the street like Gustav Havel. Because that's how Havel died, Franta Šťastný's teammate at Jawa. Havel, who wanted to do nothing but keep his mouth shut and race. He ended up being killed in Prague by some lady who stepped off a curb without looking.

Who could tell how people would die? True, among the heroes of Frank's youth, riders like Ray Amm, Keith Campbell, Renzo Pasolini, Jarno Saarinen, Bill Ivy, Dickie Dale, Les Graham, Tom Phillis, John Hartle, Bob McIntyre, and Fergus Anderson were killed on the track. Anderson, twice world champion for Moto Guzzi, died at the age of forty-seven on a works BMW while chasing John Surtees in Belgium. But Gary Hocking was killed in a car, and the great NSU rider Werner Haas, the first German world champion, retired unscathed in 1954 after four world titles, only to die in an air crash a year later. So who could tell?

Frank continued racing. In 1975 he won the WERA U.S. Championship again in Unlimited Production and in Unlimited GP. In the Canadian championships he finished second. His endurance teams also won three races for Lester Wheels and for Kawasaki, two at Nelson Ledges, and one at Summit Point, West Virginia, and came second in the 200-miler in Charlotte, North Carolina.

So he was now twice Canadian, three times American champion in two classes. Not bad. Frank always liked going deep into corners; well, he was going in deep now and was ready to go in even deeper.

There was only one problem. Many years later the Canadian racer and moto-journalist Chris Knowles would write that Mrazek was, "in many ways the father of modern-day superbike racing". Yes, very nice; but as Jana pointed out, he was also the father of two little girls by then: Sabina and Janette (or "Puppy", as everyone called her).

When were they born? Okay, they were born, Frank would reply, they were born, let's see . . . and that was the problem.

Frank loved his girls with all his heart but for him "man's love" was clearly "of man's life a thing apart". Lord Byron could have had him in mind when he wrote the line. Frank doted on his daughters, he couldn't have been a more adoring father, except he wasn't quite sure when they were born. Dates were never his strong suit. Ask Jana, he'd say. He rather thought, when pressed, that Sabi was born around 1969 and Puppy, around 1971.

In actual fact he was right, but it was a shot in the dark.

14

Smoking the Champion

Racing wasn't only racing. It didn't only happen on the track. It wasn't just cracking open the throttle, looking for that feeling, hitting the groove, or pulling that ecstatic long wheelie after the chequered flag.

Racing was everything that went with it. Even chores like loading the van, chasing spares, or trying to straighten out things that got bent were somehow part of it. Even those bits that taken separately were a pain in the neck added up to a joyous expectation. They were like the great long drumroll in the circus before the act.

More than anything, racing was all the stuff that never made it to the paddock. The tedious, dreadful stuff, the papers, the politics. All the garbage that was just a yawn or a bother, the bills, headaches, customers, suppliers, taxes. They just went whoosh. They all vanished as if by magic for those two or three days at the track. The endless nonsense of trying to get along at home or at the job, the stupid things that got everybody climbing the walls all week, were suddenly gone. A guy could take a deep breath and concentrate on things that *mattered*.

As for people, they became different. They became okay or even more than okay—well, not always, but eight times out of ten. It made sense because nobody wanted to be an asshole. The track just wasn't the right place for it. Who'd want to pull something funny when the next day he'd be going into a sweeper at one hundred miles an hour with a bunch of guys leaning on him? Or when he could be stuck without a screw or a wrench or a piece of wire and no one to borrow them from?

But it was more than that. Because it wasn't only caution or prudence. It was understanding and respect. On the track you'd be chasing those guys and they'd be chasing you. You'd be dicing without mercy, but also without disdain. There's rarely any disdain when everybody's under the same huge shadow. You pay a lot of respect. You know that those guys have put their money where their mouths are, too.

So most people were friendly. They were fun. They helped each other out when they could.

Frank remembered stuffing the H1R into the Armco one weekend at St.-Eustache, Quebec — wedging it somehow between the steel barriers so deeply that it took the marshalls three-quarters of an hour to pry it loose. He was banged up himself, though not very badly because he hadn't slid into the barrier.

Frank was always careful to have lots of ice in the cooler before going to the track. It could make the difference between racing or not racing the next day. Ordinary bangs and bruises could be cured with ice and pills as long as you put on the ice right away. If you didn't, you might not even be able to move your arm or leg within an hour. The blood would just clot in big lumps under the muscles or the skin and then that would be it. Out of action for the whole weekend, you couldn't even get into your leathers again.

But you couldn't put ice on an H1R. The thing was bent like a pretzel, including the frame. The fairing was in pieces.

Which was when Dan Sorensen said to Frank, "Don't worry, you'll be racing tomorrow."

Sorensen was a good guy, very fast, and Frank had often gone racing with him. Not as a team, but as friends. The two of them would put up a tent, and for dinner cook sausages over an open fire. Frank would bring thirty to forty pairs of European kolbasi which they'd prepare by the old camping rule of pink is dinky/brown's broiled/black's buggered. No fifty-dollar meal in a restaurant would ever taste as good, nor would a restaurant let them get away with only fifty dollars by the time they finished all the booze. Frank favoured neat vodka, though by morning he would be as sober as a judge.

"Forget it," Frank replied to Sorensen. The Kawasaki looked hopeless and Frank could hardly keep his eyes open because of the painkillers in his system. Some weekends were just a write-off, and this looked like one of them.

"Yeah, okay, go to sleep," Sorensen said. "I'll fix it for you."

He and maybe two or three other guys must have worked half the night because when Frank crawled out in the morning, all achy and shivering, the bike was sitting there. It looked terrible—taped or wired or more likely *sewn* together with needle and thread—but it was there for Frank to take to the grid, just as Sorensen had said it would be.

Sorensen was going for points in the same championship as Frank. He was in the same class, competing against him for the same trophies, the same sponsors, the same purse, the same chequered flag, yet he worked all night with some guys to make sure that Frank had a ride. And okay, Sorensen was a friend, but Frank didn't even know some of the fellows who helped him, who worked with him on the bike half the night, and they all had races the next morning, too.

People just did this at the track—true, they did it more often in the regionals than in the nationals, or in the nationals more than in Grands Prix. When they raced for the factories they could rarely do it. Not that a lot of factory racers wouldn't have wanted to do it, because it was habit, it was in the air, but their team bosses wouldn't have let them. Helping competitors was hardly part of the ambiance of the big manufacturers' boardrooms.

That's why Frank was half-glad sometimes that he wasn't racing in the international leagues. This, in St.-Eustache, was pure racing. Monza or Salzburgring were racing too—hell, they were spectacular racing—but they were also politics, and the years in Canada hadn't changed Frank's attitude to that.

At St.-Eustache Frank was obliged to get on his bike once it was sewn together, obviously, since the guys had worked on it all night. Luckily (if that was the right word) in Quebec they didn't make you tech in for a second time after a crash. If they had, the H1R could never have passed the scrutineers. The frame was bent so badly it wouldn't even track in a straight line. Still, Frank somehow managed to finish in second place—and not even *behind* Sorensen, because after fixing up Frank's hopeless wreck Sorensen himself DNF'd with some piddly thing like a fouled plug or a broken wire. But that was racing.

Racing was also the young guys. Every year new youngsters would be coming up, and you could tell right from the start that some of them were going to be hotshoes. Frank began having a few such hopefuls hanging around him because he was older. He was a kind of father-figure, and he also had a dealership and repair shop where novices could gather to bench-race or tinker.

Frank had a soft spot for them, for the good ones, like Lang Hindle, George Morin, or Steve Gervais. These future champions were only kids then. Some would do chores for Frank, run errands, act as gofers, and in turn Frank might lend them a bike to race, or set up their own motorcycles for them. When he helped Steve Gervais set up his first bike, a 250, Gervais started out with the experts at Mosport and beat them all. He just dusted them. The veteran who came second in the race finished maybe twenty-two seconds behind Gervais.

Frank wasn't a systematic teacher, a theorist, an explainer, but he could teach by example. You had to be good to begin with before you could learn anything from him. For instance, if you could keep up with him, he'd let you see his lines. He wouldn't wait for you—what was the point, at slow speeds the lines would all be different, and Frank couldn't drive too slowly anyway—but he'd go "eight-tenths" to let you see what he was doing. If you could keep Frank in sight when he was going eight-tenths on a racetrack you were worth showing some stuff. If not, forget it.

He would teach novices the finer points of racecraft. At least the things that worked for him, because everybody rides differently and what one fast guy is doing may not work for another fast guy. He would tell them how to glance around the starting grid, watch out for riders who seemed too pumped up or nervous, the people who, as Frank put it, "may go banana" on the first lap. These were the riders who could take you down, either directly or because they'd spook you into going for it too quickly on cold tires. Sometimes it was better to wait, let things settle down a bit—though sometimes you couldn't afford to wait. You had to know when you could or couldn't cruise for a lap. It was part of racecraft.

He'd show the kids how everybody got faster when following a faster guy. Not only by drafting him on a long straight—that was obvious—but just by picking up his rhythm. You could get as fast as a faster rider in front and still have something in reserve for passing. On the other hand, passing too early might only slow you down.

Frank would warn novices not to get riled by people chasing them, not to lose concentration or start craning their necks. "Track is not highway," he'd tell them. "I'm in front, okay? If someone buzzing my ass, I just go my line. I don't even move millimetre. If you're faster, if you like to pass me, *you* find the room."

But more than anything, Frank could show people how to ride in the rain. He was a natural rain rider. Even people who could pull him on a dry track couldn't get near him in the wet. It was all in one's head anyway, but the rain could leak into people's brains and short-circuit them. Wet tracks could reduce the limit in riders' minds by several seconds below what it really was, but Frank's mind was waterproof. "Forget it, you ride like dry," he'd say. "What for you worry? You *feel* when bike go wiggle."

He liked the youngsters for another reason. He was, after all, not just a trainer or father-figure but Crazy Frank. He could party with the lot of them and even outlast them, as likely as not. He felt about these young bloods a bit like the old Russian general in Prague must have felt about him back—when?—God, maybe a thousand years ago. The only thing he didn't like was when some of the kids started smoking up. That wasn't part of Frank's world. He thought that hash and pot were plain stupid. He even hated regular cigarettes; the smoke bothered him.

Booze was strictly after-hours stuff for Frank and it never got him into any trouble. The one exception was when they threw him in jail in North Carolina for running moonshine.

It was an interesting charge, especially since Frank had no idea at the time what the word meant: "Moon-shine? What this cop say to me, moon-shine? He gone cuckoo or something?"

He was with Steve Gervais when it happened. They were going to the track at Charlotte for an AMA National, the U.S. road-racing championship, involving up-and-coming hotshoes like Gary Nixon, Kenny Roberts, Richard Schlachter, Richard Chambers, and Steve McLaughlin. The promoter was offering start money to top riders, pretty rare by the mid-1970s, but even apart from the money Frank always liked the track. It had really tight banking with tremendous G-forces, more even than Daytona because of the shorter radius. The pits and paddocks were also comfortable and up-to-date, so Frank was looking forward to racing in North Carolina.

Frank and Gervais were riding in the van, pulling their bikes on a trailer. Just a few blocks from their hotel they saw a liquor store in a plaza, so they pulled in to tank up on a bit of booze. The stuff was cheap, big gallon jugs of vodka and gin, so Frank ended up buying about eight of them. Who could tell how many guys might show up for a party?

They paid the clerk, got back into the van, and were only about a block from their hotel when suddenly they were surrounded by flashing lights. They stopped, and next minute there was some cowboy tugging at the door of the van, wearing a big hat and a big Colt in an open holster, demanding Frank's driver's licence.

Frank gave it to him.

"What's in those bags?" asked the trooper, pointing to the jugs of liquor.

"Booze," replied Frank innocently. "We buy just now in store for take to hotel to drink."

The lawman of North Carolina responded to this confession by opening the door and grabbing Frank by his shirt. The next thirty minutes were a whirl of activity. At the end of it van and trailer had been hooked up to a tow truck, and Frank was behind bars in the holding cell of a police station. Gervais, who hadn't been arrested, was sitting outside on the curb. He had no credit cards or money; whatever Frank didn't carry in his pockets had stayed in the van. Gervais didn't even know the address of their hotel.

At the station the cops kept talking at Frank, but Frank could barely cope with English, let alone whatever language troopers speak in North Carolina. In fairness to the policemen, they probably didn't understand Frank much better. Finally they found some older detective who spoke a few words of what did sound more like English, and that's when the moonshine-business came up.

Apparently they were in a dry county. A person could only carry one bottle in the car without a special permit from the police. The clerk at the liquor store, as it turned out, had a deal with the cops to call them the minute a customer bought more so that the police could check him out. If he didn't have a permit, they could clap him in jail and confiscate the booze.

What happened with all that confiscated liquor the cops didn't say. Maybe they split it with the liquor-store clerk who gave them the tip. Maybe they even gave some to the judge they dug up for Frank a couple of hours later, because the old gentleman not only had pink cheeks with white stubble but could hardly sit straight when they propped him up on the bench.

By that time the race promoter had also arrived to vouch for Frank and explain the situation. The court understood, but the law was the law.

"Okay, sonny, so how much money have you got on you?" he turned to Frank.

Here Frank made an error which he realized in a second, but by then it was too late. He actually pulled out his wallet.

"Give it here, sonny."

The old judge, drunk or sober, could count with no trouble, and he counted out every penny of the five hundred dollars that Frank carried on his person. "Okay, sonny, this is your bail," he said, then handed the wallet back to Frank. He was a nice guy, he didn't take the credit cards.

The story had a happy ending, although Frank lost his bail. He wasn't going to hang around in Charlotte for a week until his trial, any more than any other tourist would have, which was the beauty of this little municipal scam. Back home Jana wouldn't let the matter rest. She wrote to the American embassy, then to the governor of North Carolina. Miracles never cease, and the governor returned Frank's money, every last cent of it, about a year and a half later.

The race itself in Charlotte did not have a happy ending, because first Frank's Superbike blew an engine in practice, and then he got a flat in the front tire of his Yamaha TZ-750. Frank was on top of the banking when it happened, full out in sixth gear, when suddenly the thing just went pshhhhhh. One second Frank was cooking, the next he was coming off the banking like a stone. He kept it upright, but nearly took two or three guys off the oval with him.

The flat taught him another valuable lesson, one of many that riders keep learning on the track even after twenty years: on a steep banking never put plastic caps on your tire valves. If they expand in the heat, the G-force can push the valve, and pshhhhhh. Use only steel caps.

Though Frank wasn't at all superstitious, he decided to sit out that AMA National in Charlotte. Richard Chambers kindly offered to help him change his tire, but Frank just shook his head. First the arrest, then the blown engine, and then the front tire: that came to *three* bad things in *three* days, right? He wasn't going to push it. Somebody was trying to tell him something.

In 1976 Frank's team won the U.S. Endurance Championship. Lester Wheels came first in all the big ones that year, including the famous twenty-four-hour at Las Vegas. They not only won it, but won it by a pretty fair margin of 147 laps.

It came about because the Lester Team—Lang Hindle, Derek Mitchell, Dennis Laidig and Crazy Frank—decided to ride through a sandstorm. Why stop because of a few tumbleweeds? They may not have been able to see the track but they could remember which way it was going, couldn't they? If the officials had hauled out the red flag they would have stopped, obviously, but nobody saw any flags so they just kept riding. What the other teams did was their business.

Winning Las Vegas wasn't painless, though, because of what had happened earlier, at the qualifier for the big race. Once again, the place was Charlotte.

The brakes on the big Kawasaki 1000 were giving the team some trouble, so everybody kept fiddling with them. They fiddled through two days of practice; then during the Sunday race they fiddled some more. Three or four mechanics fiddled, and so did Lang Hindle. Hindle wasn't only a fast guy, a born champion, but he also liked to play with the hardware. What ended up happening was that a bolt backed out of the front caliper. The mechanics thought Hindle had tightened it; Hindle thought the mechanics had tightened it, but in fact nobody had tightened it.

The bolt backed out of the caliper during Frank's shift about forty minutes before the end of the race. The team was leading, it had four laps on the runner-up team, and maybe a minute or two of dusky daylight left. The Charlotte qualifier was run in about two hours of daylight, then in about an hour of darkness. Frank was just coming off the banking, gearing down for the lefthander leading to the infield when it happened.

Frank saw his marker and grabbed the front brake. There was a big bang, followed by nothing. No more front brake. Rear brakes are of limited use in racing. Under heavy braking there's almost no weight on the rear wheel and the back end can step out at the slightest provocation. Some riders never even touch the rear brakes, but now Frank touched his, gently, since they happened to be the only brakes he had left.

The bike started skating. It skated in the direction of the wall which was coming up fast. "Concrete wall," as Frank remembered it, "with million tires around, what I hate because it's *hurt*. Holy cow. With these tires, I know it's hurt."

It was actually a "slow" accident, leaving Frank some time to think, "Holy shit, what I do? If I jump off, I destroy completely bike. So I gently put that bike—gently, well, you know, pshh, bingo, put that thing down. Let's see what happen.

"So what happen, that bike go to the wall. I roll after it, and that goddamn motorcycle hit the wall and bounce back and run me over. Ooh! It hurt. And after that I hit the wall. Bounce back just a little because I don't bouncing that good like bike."

Inadequate as Frank was as a bouncer, he did bounce back on his feet right away. He was half-knocked out, in a bit of a trance, with no thought other than getting back on the Kawasaki. They were leading the Vegas qualifier. The adrenalin was pumping. He tried lifting the motorcycle, the corner workers helped; and then somehow he was back in the seat, the marshalls pushed, and the bike fired. Frank made it into the pits.

The Kawasaki was a disaster, of course: no brake, no lights, the fairing in pieces. They taped it up as best as they could, but by then their four-lap lead was almost gone. It was also dark.

"Can you finish?" somebody asked.

"Yeah, I get on the thing here," Frank said, sounding a bit shaky. Derek Mitchell tried to hold him back, tried to get on the lightless, brakeless wreck himself, but Frank shook his head—carefully because it was still buzzing. It was his wreck, his shift, and he was going to finish it.

There was a dispute with the officials before they let him back on the track. They threatened to bring out the black flag, but everybody swore that Frank was going to ride slowly. Officials are reluctant to black-flag the leader in an important race, so finally the marshalls stood aside. By then the Lester Wheels team was no longer leading, but they were still in second place.

There was moonlight on the front straight. Frank's plan was to get behind another bike—a backmarker, anyone, as long as the guy had lights—but then he remembered. No brakes. If the guy in front grabbed a handful of *his* front brake Frank would probably ram his ass. He decided to forget it. Cruising in the dark was safer.

In fact, only the banking was really dark. The banking was black as hell but a full moon had lit up the straightaway, and suddenly Frank noticed something curious. There he was, driving slowly, eighty to ninety mph maybe, and he caught a glimpse of a thing pointing at him from the left handlebar. Could he have twisted the clutch lever in the crash? But no, the lever was where it should be.

It was his thumb. It had to be. It was his own left thumb pointing at him—but why on earth should it do that? It couldn't be broken. Surely he'd feel something if his thumb were broken, but Frank felt nothing. A little cracking maybe as he tried to move it, so he decided to stop experimenting with the thing. Nobody needed a thumb to finish.

It took another fifteen to twenty minutes, and then the chequered flag fell. Lester Wheels finished second. Frank drove to the pit, let some people grab the bike and fainted. He came to on the stretcher just as they were unloading him from the ambulance.

"Hold it," Frank said. He could remember everything clearly. There was something that had to be done because he had always made a point of doing it since Mosport, but he was still in his racing leathers without a wallet or change.

"You guys got dime?"

Somebody fished out a dime. Frank had them wheel the stretcher near a wall telephone and he gave one of the fellows a number for a collect call. When it started ringing, Frank took the receiver.

"Okay, we finished second," Frank said. He spoke casually, naturally in Czech, as he always did with Jana. "Are the girls okay? Good. Yeah, sure, everything's fine. In one piece, yeah. When will I be back? Well, I'll call you tomorrow, I must find out about the flight. Bye."

Then he handed the phone back and let the attendants wheel him into the hospital.

In the middle of Frank's third decade of road racing another revolution was about to occur in the sport. Once again, it was an assault on the ways of Europe. Once again, it had to do with a new technique and a new nation emerging on the Grand Prix scene.

The technique this time was steering. The nation was America.

To some extent the new technique was the result of technology. It came from various breakthroughs in tire and chassis engineering. It came from racing slicks which, along with new alloys and frame configurations, could absorb unheard-of cornering loads. As engines increased in power, new rubber and metal evolved to meet the challenge of putting this power to the ground.

In part, however, the technique came from the dirt. It came from the way dirt trackers have traditionally been guiding their powerful motorbikes around the loose stuff at tremendous speeds: *steering them with the rear wheel*.

This technique, easy to describe but not so easy to execute, involved "squaring off" the turn. It required a short, sharp change of direction, followed by powering out of the corner with the rear wheel sliding. As the rear drifted around, it would "steer" the front. In dirt racing the front wheel would actually point in a direction opposite to where the bike was heading.

Dirt tracking had always been a very American sport. By 1978 a young American ex-dirt tracker named Kenny Roberts would translate a modified version of dirt track drifting to the asphalt. Translate it all the way to the top. Roberts would be crowned the new king of road racing by winning the 500 cc world championship in 1978, then defending it twice through 1979 and 1980.

Nor was rear wheel-steering Kenny Roberts a lone maverick. Another young American named Randy Mamola was right behind him. From 1978 the story would be the Stars and Stripes in the world's top ten, with riders like Roberts, Mamola, Steve Baker, Pat Hannen, Eddie Lawson, and Freddie Spencer.

By 1983, for instance, the top four finishers in the 500 cc world championship were all U.S. riders: 1. The new World Champion Freddie Spencer; 2. Kenny Roberts; 3. Randy Mamola; 4. Eddie Lawson. In a way it was amusing. Only a decade earlier U.S. riders were regarded as second-rate in terms of Grand Prix racing, and now they owned the world championship. After "King Kenny's" debut, Europeans would recapture the world title in the premier class only twice. The title would go to Italians Marco Lucchinelli (1981) and Franco Uncini (1982) but Americans Mamola, Roberts, and Spencer would be right on their tails, along with New Zealand's Graeme Crosby.

By the early 1980s no rider could hope to be in the top ten in world championship racing without being able to "steer American" and drift his wheels in some turns.

However, "King Kenny" was not the first U.S. rider to win a world championship. That distinction belonged to Formula 750 World Champion Steve Baker. The Yamaha rider from Bellingham, Washington, had been a number one plate holder in the U.S. and Canada on his OW31 GP factory machine. In 1977, at the age of twenty-four, Baker became number one in the world.

First in the world—but not before coming second to Frank Mrazek.

In a way it wasn't a big deal. It was only another race. If racing is what you do, over the years you'll start in a lot of races. You'll win some, lose some, and won't remember half of them.

The race at Mosport was the 1976 Grand Prix of Canada, but even so it was just a race. Steve Baker would probably not have remembered much about it, and neither would Frank, if it had turned out as usual.

The usual way for races to turn out in those years was for Steve Baker to win them.

Baker was hot. The slight American youngster had been hot to begin with, and became simply untouchable when he started riding for the Yamaha factory team. Nobody could catch him at Mosport. Frank tried several times. He'd wiggle after Baker on his own Green Monster, but Baker just ran away from him. In the 1974 Confederation Cup race that decided the North American championship in some classes, Frank only managed fourth on his big Kawasaki behind Steve Baker, Chris Manley, and Dan Sorensen.

Frank did better the next year because he had purchased a Yamaha TZ-700 for himself, a four-cylinder two-stroke rocket which was to become a favourite privateer weapon in those years. On that bike he managed to come second to Baker. It wasn't only Frank's doing but Erv Kanemoto's, who had made special cylinders for Frank's TZ-700.

Kanemoto, then a builder-tuner for Gary Nixon, was a genius. There was no other word for it. Later he would tune Freddie Spencer into three world championships in two classes. (Just as the Australian world champion ex-racer Kel Carruthers would tune Kenny Roberts into *his* world championships. Behind every great champion, with hardly any exceptions, there is always a great tuner.)

In any event, Kanemoto had a soft spot for Frank, perhaps because Frank was already a grizzled privateer trying to chase all these speedy youngsters. He kept building top ends for

Frank and made his TZ-700 (and later his TZ-750) go like hell. However, Baker and his OW31 factory racer were faster still.

The 1976 Grand Prix at Mosport started out in the usual way. Baker had won the 250 and 500 cc classes, and was getting set to clean up in the 750 Formula One as well. That, after all, was the class in which he'd be world champion within a year. A race with Baker in it was really a race for second place. Unless the champion DNF'd the others didn't really stand a chance.

The day had been dry to begin with, but by lunch there were a few dark clouds floating in the sky. Frank kept casting hopeful glances at the gathering murk because, unlike many riders, he liked the rain. For him a wet track was a great equalizer. His friend Richard Chambers, another fast U.S. rider from Virginia, noticed Frank's prayerful glances and decided to lend him a hand. He let out a holler and broke into a rain dance. As Frank remembered it, Chambers just "started boogie, like whole bunch of Indians."

Within a few minutes the skies obediently opened up. It was coming down in buckets. Frank now had his wish, but he also had a problem. Somehow he had forgotten to put his regular tires in the van. It is impossible to race on slicks in the rain.

Chambers, whether because he felt responsible for the downpour or because he didn't want to throw away *his* brand new TZ-700 on a wet track, elected to sit out the race and offered to lend Frank his rain tires. The riders were already being called to the starting grid when half a dozen guys were still grappling to help Frank and Chambers swap wheels on their Yamahas.

The officials actually delayed the start for a couple of minutes to let Frank make the grid. Then the flag dropped. Baker, who would later describe himself as "probably not the best rain rider in the world", nevertheless shot immediately out in front.

Frank started out cautiously. He had never run on rain tires before. In the past he had simply used his regular tires for the wet. The huge Goodyears, as wide as slicks but deeply grooved, were then a new development in tire technology. "Holy cow, how I can race that thing here," Frank kept muttering as he always did on the track. He muttered like some concert pianists or conductors do, issuing various bits of advice or command to himself. It was a good thing that under the chinguard no one could read his lips.

After a few corners Frank began to like the Goodyears. Those things were *sticking*. He started from the back as usual and now he had to slither through the field. He kept slithering, and before long he found himself on Baker's tail.

It was a fifteen-lap event, and Frank wanted to win it. Not because it was Steve Baker. At that moment the idea of getting by Baker the factory guy, Baker the reigning hotshot, had nothing to do with it. He was just a rider like any other and Frank wanted to pass him. So would Baker want to pass Frank, whether it was a big deal to pass Frank or not. This was racing.

Still, it came as a surprise to Frank when he first got by the young American on the uphill Mario Andretti Straightaway where he ought not to have been able to pass a factory OW31 with his TZ 750, not even a TZ 750 tuned by Erv Kanemoto. But then it dawned on him: Steve Baker couldn't get all his horsepower to the ground.

Baker's bike was around sixty pounds lighter than Frank's. Baker himself was forty or fifty pounds lighter. Normally that's an advantage, but now Baker's wheel must have been spinning going uphill. He must have been skating all over in the wet.

They started swapping the lead. Sometimes they changed leads four or five times a lap. It was close racing, and in the covered front stands along pit row the spectators were going wild. At Mosport Baker was king. King? For the fans he was God—and now Frank was chasing him.

They were going downhill into The Chute side by side on the last lap, Frank on the left and Baker on the right. They were close to the halfway mark with about a mile and a third to go before the finish line. Moss' Corner was coming up, the famous double-apex hairpin, the slowest turn of the circuit.

If Frank was going to make his move it had to be soon. He had to pull something on Baker and it had to be something smart. This wasn't playing with a novice. The next turn was a very tight righthander. The American had the inside line, and it gave Frank an idea.

He started braking a fraction earlier than usual. Baker was watching him and started braking as well. The second Frank saw Baker's hand move on the brake lever, he cracked his throttle.

The TZ-750's front wheel shot past Baker on the outside. Instinctively Baker cracked *his* throttle to keep up. Frank let him go and immediately went back to braking. He was sitting pretty. He was set up on the outside, ready to go high, wide and handsome, while his young rival had the tight, inside line. They were now in Corner 5/A, which meant that 5/B, the wicked thirty-four-foot-radius second apex, was coming up fast.

Baker grabbed his own brake a split second later. It was too late. He had been suckered. He was now forced to start a tight turn under heavy braking in the wet.

Even at best Baker would have had to go wide at the exit and Frank could drive under him—and the best didn't happen. The OW31's front tire let go under the double load of braking and cornering on the soaked asphalt—pssst, just like that—and Baker was down. Frank squirted past the sliding rider on the outside and the next second was gone behind a curtain of spray.

It wasn't much of a crash. The rest of the field hadn't even appeared in The Chute when Baker was back in the saddle chasing furiously after Frank, but by then Frank was going through The Esses on his way to the chequered flag and victory. Baker finished second.

The Mosport crowd went wild. Richard Chambers, whether because he was Frank's pal or because it was his rain tires that won the race, was going pretty wild himself. Nobody could remember anyone beating Steve Baker at Mosport before.

Baker was upset. This was natural enough; it was more curious that Frank himself felt a bit upset by nicking the paint on what could have been an absolutely flawless, three-victory day for an outstanding young champion. He flapped around Baker apologetically in the pits for a while. Maybe it was perverse, but that's how Frank felt after the race. Mind you, during the race he couldn't have cared less.

In any event, smoking the champion at Mosport didn't alter the fact that Steve Baker was a twenty-three-year-old with a brilliant future, while Frank had just turned forty with no more than a respectable past. Not that Frank's own future was dim, because his greatest championship years lay just ahead.

In 1978, riding a Minolta-Honda CBX six-cylinder Superbike, Frank won the Superbike Championship of North America. He came first in twelve races that year aboard Honda's virtually untamable monster. There weren't many other riders who could wrestle the CBX into submission. In spite of its tremendous power it was hardly a bike for the race track.

The 1978 Superstar Ballot, a selection by WERA racers from among competitors around the world, chose Frank as International Superbike Rider of the Year. On the same ballot, returning Isle of Man TT winner Mike Hailwood was chosen as International Production Rider, and newly crowned world champion Kenny Roberts as International Professional Grand Prix Rider of the Year.

These were the three international superstars of 1978 by the vote of their peers in America: Hailwood, Roberts and Mrazek. "Mike the Bike" Hailwood and "King Kenny" Roberts. For "Crazy Frank" Mrazek it would have been difficult to be in more august company.

Frank felt honoured. He was proud and happy, but the questions that immediately ran through his mind had to do with next season. What was the Holy Edge going to be in 1979? Which sponsor might treat him to a more competitive ride?

Jana had a completely different question.

In 1978 Frank had turned forty-two. He had been selected International Superbike Rider of the Year. He had beaten Steve Baker. In addition to amateur titles, he had won four U.S., two Canadian, and a North American championship.

What else did he expect to win? Who else was he proposing to beat? Was he waiting to beat some people who were not yet born?

Normal athletes went as far as they could, then retired. They became mothers, fathers, trainers, or business people.

Frank was already crippled. Was he now waiting to be killed? Or was he just waiting for people to start beating *him*?

For Frank the answers were simple. Jana was right: he was a bit crippled. He couldn't walk too well. He did have a plastic joint in his ankle and, since Charlotte, a metal plate with a few screws in his thumb. But he was definitely alive and, hell, people hadn't started beating him yet.

There was one other thing. Frank liked winning and hated losing, true, but he also had something in common with back-markers who had never won anything. It was pure passion. Honoured guest or Lazarus, feeding on meat or crumbs, he had to be at the feast.

15

THE BOOGIEMAN

On the face of it Frank was right because the following year he repeated the performance. He became North American Superbike Champion for the second time. Altogether he won twenty-two races in three different classes in 1979.

Business, however, wasn't too good. In fact, Frank closed down his Cycle Centre the same year. You couldn't make any money selling and servicing motorbikes anymore; at least Frank couldn't. It used to be fine in the beginning: one year Frank was even chosen best Kawasaki dealer in Canada and received a truck as a bonus from the company. But then all motorcycle manufacturers became greedy; they overproduced and started forcing dealers to carry big inventories. Everything became politics. Frank had to commit himself to crazy projections to qualify as a dealer. He had to live with whatever the number-crunchers dreamed up at head office, with floor plans, twenty-five-percent financing charges, and unsold bikes in crates all over the place. It cost Frank over $60,000 in losses in 1978 alone.

So he just said screw that, and stopped doing it. Didn't go

bankrupt like some other small dealers, just sold everything. Didn't even lose money on it, except he was now sitting there without a business and without a job. So he looked around, and started building performance cars from kits.

That was okay for a while: Kelmark cars, all kinds of different rear engines in a VW chassis, with nice bodywork. Looked a bit like Ferraris but much cheaper. The kits didn't sell because they were too expensive, but the finished cars sold okay at around $30,000.

Then the government jumped on it. It was politics as usual. The bureaucrats were looking for more money to spend on themselves and came out with some crazy new budget in the spring of 1980. Rich people who had bought the cars for toys couldn't write them off anymore. By Christmas Frank was left with only a single order, so he had to get out again, fire his mechanics, good-bye.

The government—well, it wasn't as bad as Czechoslovakia, but close. Fine; let them gouge taxes from half the population and pay it out in welfare and unemployment to the other half if they thought it was so smart. Frank wasn't going to argue. He had more important things to think about.

Like next season's racing.

Some people were always looking for a perfect ride, and if they couldn't get it they'd rather not race at all. Frank was the opposite. He'd ride just about anything if somebody was willing to sponsor him.

Lester Wheels? Great. Endurance for BMW? Fine. A short-stroke Ducati for Keith Harte, the famous Canadian Ducati specialist? Marvellous. And when no one offered him a competitive ride, Frank was willing to buy his own. He had bought the Yamaha TZ-700 and 750 two-strokes with his own money. He bought himself the huge CBX Honda although Honda's Tom Faulds helped Frank with discount services and parts. And finally Frank also bought the titanium-magnesium miracle, Steve Baker's ex-Yamaha factory racing bike, the 750 cc OW31.

Yamaha had made only three of these machines. Naturally

they were not for sale to anyone except select privateers, and the price was an astronomical $20,000. Frank said fine, provided the bike was in the same condition as when Baker raced it. However, what he received from tuner Bob Work turned out to have broken cases, a broken crankshaft, worn cylinders, and pistons: in other words, a pile of junk. Frank gathered it all up and dumped it on the president's desk: Look! So Yamaha agreed to replace everything and gave it to Frank free of charge for two years.

Frank tried out his dream-bike in a few races, then took it to Daytona. He took it on a Michelin-sponsored trip to ride the tire company's new product on the world-famous circuit. The plan was to contest the Daytona 200, North America's most prestigious motorcycle race, for the Michelin team.

It was a very professional team, with Grand Prix racer John Long and Venezuelan ex-world champion Johnny Cecotto as the other two riders. Frank's tuner was Alec Maize, and Michelin sent a tire technician to look after the rubber. He turned out to be a fellow Czech, Jan Orada, and he and Frank hit it off splendidly.

It was the motorcycles' rear wheels and Michelin's new experimental tires that did not hit it off at Daytona. During practice Long and Cecotto had the same problems with the tires that Frank did. They wore out too quickly. The rear skins had to be changed after every few laps, and there was no way that the tires would last the two hundred torturous miles of the big race. Orada was virtually sitting on the telex machine all week exchanging messages with France, until finally a shipment of new tires arrived for the team on Sunday, the very morning of the race.

The new slicks might have been perfect—if they had fit. They were just a touch too high and wide for the rear wheels. The bikes were sitting on them in a nose-down attitude like dive-bombers, but the main problem was that the rubber was touching the driving chains. The huge slicks couldn't quite squeeze into the space between the swinging arms in the rear.

The race was about to start, so the team decided to do

something reckless. (Well, "reckless" is debatable; other teams had done the same thing at other times.) Frank would later say that he "didn't feel good" about it. Probably nobody did, but they did it all the same. They cut grooves in the walls of the tires on both sides for clearance, with a little more on the left side for the chains.

It was a bit crazy. True, tires are generally over-engineered and may not need all the rubber around the carcass for everyday use, but they often need it on a racetrack. Any racetrack, let alone Daytona. By then motorcycles could reach speeds of over 180 mph or nearly 300 km/h on the banked ovals of the great Florida track. In a fifty-seven-lap, two-hundred-mile race this added up to one hell of a stress, equal to the most gruelling tests ever devised for men or material. Daytona was not an ideal place for adjusting tires with a knife. But the guys had come to race, and if they hadn't cut the tires they couldn't have raced; so they cut them.

Daytona 200 starts are in two waves of forty riders each, gridded according to their qualifying times. It is a significant achievement to qualify for a first-wave start at Daytona. In fact, it's an achievement just to qualify, because there are often one hundred to one hundred and twenty aspirants for the eighty spots on the grid. (Every aspirant is a hotshoe. In 1988, for instance, twenty-four-year-old Arpad Harmati who *failed* to qualify by finishing eighty-first, had been East European champion in 250 cc twice.)

First-wave starters form an exclusive little club. They're probably the fastest forty guys in the world that season.

All three riders of the Michelin team made the first wave. However, Johnny Cecotto cooked his clutch right at the start and he was out; then in less than a lap John Long was also out with some mechanical problem. No doubt both of them felt unlucky at the time, though it may well have saved their lives.

Frank was feeling lucky. The OW31 was flying, and he completed the first three laps with no trouble. On the fourth lap he was just hitting maximum speed coming off the banking onto the front straight before the pit-lane stands. He was probably moving at around 180 mph. He had not yet started braking for Corner 1 which leads into the infield, when something happened.

Frank couldn't figure it out. "Holy cow, this is different angle," he remembered thinking. He had been glancing at the face of the tach but suddenly he saw it in an entirely novel perspective. It was a top view, from pretty high up, as if someone had taken the instrument from the fairing and planted it on the ground. This must have been the moment when Frank was flying over the handlebars.

Frank crashed on Daytona's front straight when his macerated rear slick blasted off the rim. He crashed at a speed at which few people ever crash, probably around twice the speed of free fall. Jumping out of a plane without a parachute would have been a low-speed crash in comparison. The impact, of course, was not nearly as blunt because Frank did not smash directly into the ground. His body's momentum carried him forward and he glided, quite literally, first skimming and eventually sliding, rolling, and tumbling along the pavement. He scrubbed off a lot of speed before he hit the sand past Corner 1, and did not crash into the straw-bales until he had nearly stopped. As the friction was burning off leather from his back and his gloves, his body was trailing white ribbons of smoke, probably for about two hundred yards, until it disappeared from the spectators' view in a great explosion of straw and sand.

Frank woke up two days later on a narrow hospital bed with both his hands dangling in pails of ice water. They were infected, bluish-yellow, with bits of straw and sand still sticking to the raw flesh. Still, considering the speed, it hadn't been such a bad crash.

Nine broken ribs, five fractures on the left arm, four fractures on the right, broken collarbone, second-degree burns on his back and his hands. And a concussion, obviously. But lucky—much luckier than at Mosport. A far worse spill, but with no crippling injuries. Naturally, if he had crashed into anything at this speed, a guardrail or a wall, he would have been pulverized. Disintegrated. They'd have had to shovel him into a bag. But luckily he had hit nothing. His survival was a tribute to the role of fortune in human affairs, as well as to the design of the track at Daytona.

The OW31 was a write-off. Caught fire, the works. Not as much as a screw could be salvaged from it.

It was after his crash at Daytona that Frank started thinking seriously about retiring. Not right away, though. For one thing, his injuries were healing quickly, and within three or four months he was feeling fine. For another, retiring after a big crash was like being beaten into submission, being cowed or frightened by fate, and it was something a guy just couldn't do. He'd retire, sure; but in his own good time.

Not that Frank viewed himself as a person who could not be frightened. On the contrary, he always made a point of saying that when he and Jana were crossing the border from Yugoslavia to Austria he was so scared that he nearly let go in his pants. Frank would also admit to being uneasy about having to dispute things with his wife. For instance, he didn't enjoy facing her after Daytona this time. As he put it, "I go scare from Jana, because, oh, she hate my crashing. When she get upset, she screaming, yelling like tiger."

Frank was scared of many things; it just so happened that he was not scared of the track, and he didn't have much use for people who were. If they stayed away altogether that was fine. Everybody couldn't be a racer, that was obvious. What he didn't like was people who came to the track to make a fuss, to whine and bitch, to make rules about this or that, or to complain whenever they thought that someone had passed them too closely.

He had no use for people who really, seriously thought that he was crazy. He was proud of his nicknames—Crazy Frank, the Bouncing Czech, the old Boogieman, and so forth—but they also surprised him. Why was he crazy? So he didn't worry, he was having fun—so what? You worry about things; okay, you stay home. You race, you race. What was so crazy about coming to a racetrack to race? Wasn't that what racetracks were supposed to be about?

What kind of fool would start in a motorcycle race to be careful? It was like sitting in on a poker game to save money.

Besides, Frank was poor. Maybe a rich guy could afford to be sane. He could ride smartly and take no chances. He could be careful and slow, he could go to sleep on the track and ask people to wake him up after the race. Anybody could afford to be slow who could pay for his own racing. He could go on having fun as long as he liked.

But a poor guy had to go crazy. He had to be fast. He couldn't pay for himself, and who was going to pay for a slow guy?

Throughout 1980 and 1981 people were still paying for Frank to race, so he continued racing. Paying—well, nobody paid *him*, but they sponsored him, picked up at least part of his racing bills, and he could keep his winnings. They wanted him on a Unitour-sponsored Canadian team both years to contest the Grands Prix of Cuba, Jamaica, and Guyana. So Frank went, naturally, because it was a great chance to race in the winter. He started in twelve of those minor Caribbean Grands Prix in the 250, 350 and 500 classes, won six of them and came second in the other six. For good measure, during those two years he also won two Grands Prix in Canada and came first in the Victoria Day Sprints at Mosport.

And then he retired.

The first winter was no problem. Frank was used to not racing, or racing only rarely, in winter. The spring and summer of 1982 were beginning to be a problem, not so much of not racing, but of not having any *plans* to race. Not even to bench-race, because Frank stayed away from people who had anything to do with racing, much as a reformed smoker shuns his smoking friends. But it was still not too bad, because he was trying to build up a new business anyway, doing propane conversions for taxis and limousines. He also rented some space, put in tanks, sold propane gas to taxis. He was keeping his business open twenty-four hours, coping with hold-ups, griping customers, thieving employees, shortages, politics, the lot.

He was acting his age. Behaving like a responsible citizen of forty-six. It was fine. Nothing had gone out of his life, except maybe the joy.

He was also taking stock. Not too systematically—grave introspection wasn't one of Frank's shortcomings—but in fleeting bits and pieces. What had racing ever given him?

If figure skating had disappointed Jana in the end because she didn't get to the very top in the world, well, Frank didn't even get halfway. Jana had made fourth in the world, while Frank never even started in a world championship race, except back in Brno, because North American or Caribbean Grands Prix were not part of the world championship then.

Frank had never entered the Isle of Man TT. Except for the old Masec, his home track, he had never done a single lap on any of the renowned circuits of Europe.

And the truth was, the circuits of Europe were the big leagues. They were the big leagues still, even if they were ruled by American riders on Japanese machines by now.

He did well in North American Superbike racing, yes. He did well in endurance and production racing. He had won many national titles. He had done honourably, even made it to the top of the minor leagues—because that's what they were, the minor leagues. There was no getting around it. He had never played in the majors, unlike some of his former teammates or friends, a Mike Duff or an Yvon DuHamel.

Yet he could pass guys who had made it to the top in the big leagues. Did he not pass Jim Redman in Brno? Did he not smoke even Steve Baker once, the world champion?

Never mind world titles. He was now retired without even winning a big international race like the Daytona 200. Or *any* race at Daytona. Sure, getting the first topless national championship at Las Vegas was fun—a showgirl letting it all hang out as she handed over the trophy for the 1976 U.S. Endurance title—but it wasn't Daytona.

So this was it: a basement full of cups and plaques, a garage full of spare parts, and a few fading newspaper articles and photographs. As for the cost—well, pieces of plastic and metal scattered around in his body. No major bone unbroken, except, thank God, his spine. Most of them fractured three or four times, from his nose to his toes. All of which hadn't mattered at the time and didn't matter now, because racing was the most astounding, exhilarating, exciting, and joyful thing on earth.

Except Frank wasn't racing. He was retired.

This was the problem, not the cost. The cost didn't matter, and neither did the lack of rewards. Racing was its own reward, at any time, in any league, at any track. Except no one could reap it unless he was doing it, and Frank wasn't doing it anymore. He had retired, and that's all there was to it. From now on he could only pay the price. All he would be left with was the price, paid in installments whenever he had to stick his ankle in a whirlpool bath to ease the pain. Racing was "have fun, pay later"; and now it was later and it would stay later forever more.

Nineteen-eighty-three was a grim year. Frank tried to concentrate on the propane business. The propane business was fascinating, but a guy could concentrate on it only so much. By the end of the year Frank was "going banana". On good days. On bad days, he thought he was "going cuckoo".

In the end what saved Frank from going any kind of tropical fruit or bird, singular or plural, was his daughters. Because what Sabi, fifteen, and Puppy, thirteen, did in the spring of 1984 was to prevail upon their father to take them to a motorcycle race.

It was interesting. Sabi and Puppy had been to the track once or twice as toddlers, but they were much too young then. By 1984 they could remember nothing about it. Jana had stopped going to the track, and Frank on his own couldn't look after little children while he was competing. So all the girls knew as they were growing up was that their father would disappear on weekends to do something that their mother seemed unhappy about, and that he'd return within a few days, generally under his own steam, but once in a while in casts or bandages. They knew that what he did was called "racing" and they could see the growing collection of glittering trophies in the basement. That was about all that they knew.

Naturally, they became curious.

Frank was one of those people who have a good relationship with their children more or less automatically, without having to do much about it. He certainly never courted his daughters or sought their favour. It wasn't the European way. It always amazed Frank to see parents in North America letting their kids walk all over them, looking for their affection. He was just himself with the girls: neither too stern nor too lenient, just normal. He wasn't trying to be a pal or a nursemaid. If he was fun to be with it was because he was fun to be with, not because he was making an effort.

Frank loved Sabi and Puppy because they were his girls and he took it for granted that they'd love him because he was their father—and so they did. They even worshipped him a little, and now they wanted to see what it was that he had been doing all his life.

The girls kept at Frank for months. He genuinely resisted them, not just in self-defence, but because he was not a spectator by nature. He'd much rather do things than watch somebody else do them. It was for the sake of family peace that he finally decided to take the girls to a track nearby. What the hell, maybe it wasn't going to be too awkward or painful.

Jana didn't object. Now that the girls were teenagers there seemed to be no harm in letting them watch a race on a pleasant weekend in May. She didn't go with them, though. Maybe things would have turned out differently if she had, but she wasn't interested and she was busy.

Frank and the girls piled into the car and drove to Shannonville, Ontario, about one hundred miles east of Toronto. They walked through the gate and just about the first person they bumped into was Dan Sorensen. He had come to the track to run his old 750 Norton in the vintage races.

Frank didn't know much about vintage racing then, though

it was no longer a novelty by 1984. A number of enthusiasts in England, in North America, and in some European countries couldn't let go of the golden age. They formed historic racing associations, won the approval of national sanctioning bodies, and talked the promoters of modern racing events into letting them run a few classic cards on the program.

Soon the great British, Italian, German, and American machines of road racing's historic past, along with some earlier Japanese models, thundered to life again; and the spectators loved them. These Nortons, BSAs, Matchlesses, Ducatis, Moto Guzzis, BMWs, Harleys, Triumphs, Velocettes, Greeves, Laverdas, or Bultacos sounded and smelled like real racing bikes. The latest Oriental weapons may have been "trick"; they were certainly smooth and incredibly fast; but the old machines were pure romance.

They weren't all that slow, either, many of them. Sometimes vintage events provided fans with the closest racing of the day.

At any rate, Sorensen wasted no time. "Hey, Frank," he said, "we're all retired folk here. Why don't you start racing vintage with us?"

Frank shrugged. "I have no bike," he said.

"No problem," replied Sorensen. "I have two. I've got another Norton right here. Why don't you take it out today?"

"No," said Frank, in what he hoped was a firm and decisive voice. Then he paused, took a deep breath, and added, "Because, you know . . . Naw."

Sorensen changed tack. "I guess you're right," he said. "You let your licence lapse. You weren't here for practice."

Frank stirred. "Well," he said, "I can talk to referee."

"And you have no helmet or leathers."

"I can borrow," said Frank, and when he saw Sorensen grinning, he added, "What I can do? You twist my arm."

By then it was almost noon and the vintage race was the first event after lunch. Frank sought out the official in charge. "Hey, can I go race here?" he asked.

"But you have no bike, Frank."

"Sorensen give me bike."

"You've had no practice."

"Well," said Frank modestly, "I know track here a little."

The official laughed. "We can't grid you, Frank, unless you want to sit in the back."

"Oh sure," said Frank. "I go sit in back."

Thirty minutes later he sat on the back of the starting grid in somebody's ill-fitting leathers. There were fifteen or twenty riders ahead of him. Fifteen seconds after the flag he was going into Corner 2, groping with his right foot because it had slipped his mind that on Nortons the rear brake was on the left. Then he decided to forget about brakes because the old British bike was so slow anyway. It handled a treat, though, and Frank was beginning to enjoy himself. Ten minutes later he saw the chequered flag.

He glanced over his shoulder. The nearest bike was just coming out of Corner 6 on the Nelson short-track, about four seconds behind him. There was certainly no one in front.

Sabi and Puppy were jumping up and down. "Well," Frank explained to them as he took off his helmet, "that's what I used to do." As soon as he said it, another thought occurred to him. This wasn't only what he used to do. This was what he *was*.

He was a racer. When he was racing or planning to race he was alive. When he wasn't, he was dead. One day he was going to be dead anyway, sure; but what was the point of being dead in advance?

"What the hell," he said. "I still remember how. What do you think, girls? Maybe I go boogie a little more."

To be on the safe side, though, he didn't take his latest trophy through the front door. There was no point in coming out with things too suddenly. He told Sabi and Puppy to keep quiet, casually kissed Jana, then hid the handsome wooden plaque under the stairs.

16

A MATTER OF LUCK

People enjoy only the degree of freedom
that their audacity conquers from fear.
HENRI BOYLE STENDHAL

The racing world that Frank entered again at the age of forty-eight was different. It was better, and it was worse.

What made it better was some of the people. For a start, vintage racers included people who were Frank's own age, or at any rate only five or ten rather than twenty-five or thirty years younger. It even included some who were older, because racers in their fifties were not uncommon and a few stubborn souls in their sixties and seventies were still wheeling their historic machines out on the track.

In modern racing Frank would have had to compete with kids who had not yet been born when he arrived in Canada. Including some awfully fast kids. Satisfying as it was to dice with them once in a while, it was also humiliating. "Hey, Frank, when are you gonna pack it in?" young bloods kept yelling at him in the paddock, and there might be admiration in their voices, yes, but also a little derision. For someone Frank's age still hustling out there was somehow—well, not really graceful. It was being a kind of Don Quixote, a crazy old knight tilting at windmills.

There was none of that in vintage racing, even from the youngest competitors. Classic racing also included some kids, often the children of old-timers who were still competing themselves. But there was no mocking, not even of a good-natured kind. Fellow competitors, officials, spectators looked on in respectful silence. Old men belonged. They belonged on these old machines.

Vintage racing started in Britain way back around the mid-fifties, but it took until the early 1980s for it to become popular in other countries. The late British vintage enthusiast John Griffith was instrumental in organizing a Classic Racing Motor Cycle Club (CRMCC) section, growing out of the Vintage Motor Cycle Club (VMCC) in England around the late sixties. Eventually both clubs, along with other groups in Scotland and Ireland, began holding enthusiastic and surprisingly competitive meets.

They were not just nostalgia rallies for old-timers. They were proper races, with classic riders on classic machines on some of the classic circuits of Britain. Spectators soon started flocking to them. Perhaps it was a bulge of war-babies hanging on to their memories, perhaps it was the intrinsic appeal of solid, pre-plastic industrial art in a micro-chip and liquid crystal age, but whatever it was, vintage racing caught on.

Across the Atlantic, American enthusiasts Bob Coy and Allan Adie established the United States Classic Racing Association (USCRA), while Robert Ianucci, founder of Team Obsolete, the most prominent as well as most aptly named vintage racing team in America, helped to set up what eventually became the American Historic Racing Motorcycle Association (AHRMA) under Dick Mann and Gary L. Winn. In the south the California Vintage Racing Group (CVRG) came into being while in the north Doug and Manzi Warwick, Dave Hughes and other enthusiasts started the Vintage Road Racing Association (VRRA) of Canada.

Sanctioned by national organizations, including the powerful American Motorcyclist Association (AMA) in the United States, vintage clubs began organizing championship series and competing with each other. Soon race promoters were including vintage events in their regular programs. In 1983 the annual Am-Can Vintage Challenge series was born, held at various major tracks in the United States and Canada, such as Mosport or the famous circuit at Loudon, New Hampshire. Vintage racing was becoming glamorous. A historic round became part of the week's races even at Daytona, the Wimbledon of motorcycling, the richest of all American events.

Vintage racing quickly developed its own stars. They included some great champions of yore returning to the track for one more go, often on their original machines. At the Old Timers' Grand Prix of 1987, for instance, stellar names like John Surtees and Luigi Taveri turned up at Austria's Salzburgring. (Taveri was riding his old Honda 250-4, and Surtees, an ex-Hailwood/Saarinen Benelli 500-4.) Hugh Anderson himself appeared on a Matchless G50 at Britain's Donington Park for the 1986 Classic Race of the Year. But in addition to former world champions, vintage racing kept in the limelight dozens of long-time hotshoes like Americans Dave Roper, Pete Johnson, Kurt Liebmann, or Pat Conroy, along with such Canadian aces as Ken Hodge, Dan Sorensen, Paul Bowyer, or Gary McCaw. It attracted many fast riders of a younger set like Dave Pither (Britain), Paul "The Rain King" MacMillan (Canada), or Todd Henning (U.S.), and many others. The starting grids in some classes would be crowded enough to call for two-wave starts at major vintage events.

Frank took his place in this crowd as quickly as he could lay his hands on a vintage machine. He settled on a 1967 450 cc Honda twin, partly because it was fairly competitive and partly because it was cheap. Bored out to just under 500 cc, the old Honda ran fine and in Canada Frank pretty soon owned the 500 cc class with it. He won three races in 1984, fourteen in 1985, and nineteen in 1986. In the United States, though, the Honda was having trouble keeping up with Team Obsolete's super-fast Matchless G50s, especially as ridden by Dave Roper.

It was for this reason that while this new racing world was better for Frank, it was also worse. The people were great—but he was totally, completely at the end of his resources. He could afford nothing: no faster bikes, no spares, no crew, no tuners. Except for a single vintage class in Canada, he was no longer competitive.

Frank was racking his brains. True, he had never really parleyed his championships into anything. He had always remained an amateur, racing for the love of racing. Often he had to purchase his own bikes. He had never made any money, he had stayed a bedraggled privateer all along; but at least he used to have sponsors. Even in the worst years somebody would treat him to a ride.

But in vintage competition sponsors were virtually nonexistent. Most makers of the classic marques were long gone, and the ones still in business were naturally pushing their latest models. Honda wasn't crazy to spend money promoting twenty-year-old twins when they had warehouses brimming with state-of-the-art multis.

Manufacturers of after-market products also concentrated on modern machinery. That's where the purses and advertising funds were going. The money, what there was of it, was in Superbikes and maybe in the production classes (in addition, of course, to the World Championship series where most of the money had traditionally been). In spite of classic racing's growing popularity, it was just a hobby. It was either a rich man's toy or a poor man's sacrifice, and Frank had nothing left to sacrifice anymore.

The propane business had turned out to be more trouble than it was worth and Frank ended up selling it to his suppliers. He didn't lose on it, but he certainly couldn't afford to retire. He accepted various jobs—repair-shop foreman, gas-station manager—while he was looking around for some new business. At his age, hobbling on a plastic ankle, he was glad he could get a job at all.

Jana had lost all sympathy for racing by now—not that she had had much left after Mosport, Nelson Ledges, Charlotte, and Daytona. Frank couldn't blame her. No woman could feel any other way. In Frank's own view most wives wouldn't have put up with a husband like him for ten minutes.

He had been selfish all his life. Living at his side, what did Jana have to look forward to? What was her security? On weekends the best she could hope was that Frank would come home in his van and not in an ambulance.

It wasn't a question of money—and yet it was. They had always lived simply, okay? Money was never important to them and that's why they didn't have any, but now it would have been nice to be able to point to something for Jana: look, whatever happens, you have *this*. A nest egg, anything, instead of just a black cloud of worry and poverty.

Except Frank also knew that if he quit racing he'd be gone anyway. Gone, pffft, like the snap of a finger, that's it. His two years' retirement had proved that to him. He loved his wife and he loved his daughters but whatever life was, it was at the racetrack: he could never find it anyplace else. Nor could he wait, because there was nothing to wait for. For him this was it, right now: the blue flag with the diagonal white stripe, the last lap.

For a short while Frank still disguised what he was doing. He told Jana he was only helping out some other guys at the track. He kept hiding the new trophies under the stairs with the giggling connivance of Puppy and Sabi, who started going with him to the races, wiping his visor, handing him tools, fetching ice, pushing and loading the bike. His daughters were Frank's crew now. They even brought along girlfriends from school and they were having a whale of a time. But hiding things was silly because it didn't take Jana too long to figure out what was going on, and Frank wasn't good at pretending anyway; it was too much like politics. So he just put his head down, weathered the storm, and kept racing.

In 1986 Frank bought himself a second bike. He had no other choice. Vintage racing was divided into various periods up to 1972, and he had been pushing his 500 cc Honda, a Period One class machine, against 750 cc bikes in the newer Period Two Supervintage Heavyweight class. Of course, he might have entered only one class, but one race a weekend was hardly worth unloading the bike for. Nobody could keep up his proficiency riding in only one race, so Frank bought himself an ex-Paul Smart machine, an original factory Triumph triple, and had it shipped over from Britain.

It cost him the last of his money, but it was worth it. It was even an investment, an authentic original of old Smartie's, one of the founding fathers of the modern racing style. Collectors would pay a lot of money for such a machine; or so Paul Smart's nephew, who had sold Frank the bike, assured him. Whatever it was worth the three-cylinder bike was going like a rocket, except it was leaking oil all over the place, was the devil to start, needed a lot of work, and Frank had nothing with which to pay for parts or tuners.

Frank was stubborn, so he kept thinking and thinking. There were a few well-to-do enthusiasts who owned classic racing bikes. These fellows sometimes ran teams, like Robert Ianucci's Team Obsolete or Tom Faulds's and Tom McGill's Ecurie Classic, and they gave rides to other racers. The trouble was that they, too, seemed to be looking for young hotshoes, just like the factories. Everybody wanted young guys. Frank couldn't blame them; maybe he would have done the same thing in their place, but it was a pity.

A pity; because Frank felt that, on equal machines, he could still give the young guys a run for their money. How could he prove it, though? He would need a full-time mechanic to keep the Triumph running and he couldn't prove it on the Honda.

There was one class that was a kind of no-man's-land between classic and modern technology: a stylish, intriguing class, called the Battle of the Twins. Some of the vintage guys rode BOTT, and so did a lot of other people. A twin-class motorcycle had to have two cylinders, obviously, which was the essential requirement. BOTT races were also split into sub-classes of lightweights, heavyweights, GP, modified, amateurs, professionals, and so on, sometimes running on the track together but scored separately.

Many BOTT bikes were classic brands. It was a class in which Ducatis, Moto Guzzis, BMWs, Harleys, and Nortons still held their own. They were not necessarily vintage models—some had just come off the assembly line—but they were vintage designs, sometimes radically updated. Not too many makers continued manufacturing high-performance twins, although (especially among the lightweights) there were a few Japanese marques. The BOTT was the last class of motorcycle racing in which Europe and America could still compete with the Orient on equal terms.

The Battle of the Twins was a modern class like all others, though with sparser entries and somewhat meagre purses. It was a class for people with refined tastes, both among spectators and competitors: a sort of high-brow item on the menu for motorsport's connoisseurs and eggheads. It was a classy class. Winning the Daytona Pro Twins GP fifty-miler had as great a cachet as anything. In fact, after the Daytona 200, it was probably North America's most coveted motorcycle racing title in the view of many people.

Frank always thought that it was one Daytona title at which he could still have a shot if someone gave him a competitive bike. Just in case, he started racing BOTT on his 500 cc Honda in the Eastern Canada Challenge series. After all, the old Honda was eligible: it had two cylinders.

Though the Honda couldn't quite win the BOTT championship in Canada, Frank did win several BOTT races on it. He kept chasing the 750 cc Nortons of old-timers Ken Hodge, Dan Sorensen, and Paul Bowyer, to say nothing of such young hotshoes as Terry Spiegelberg on his 750 cc Ducati, Brian Morrow, Pat Barnes, or Sandy Mitchell on their modern Kawasakis, and eventual Canadian BOTT champion Larry Strung on his 1,000 cc BMW. Even if the young guys could beat him on their bigger or faster bikes, Frank made them work for it, and sometimes he wouldn't let them beat him at all. He treated himself to some BOTT victories for his fiftieth and then his fifty-first birthdays.

So it was fun, but it was also a grim struggle and Frank did wonder about it more and more often. For instance, his plastic left ankle wasn't getting any better. It didn't give him enough mobility to raise his foot out of harm's way on the footpeg, so in some left-handers the tarmac would rub its way through his boots to his toes. Frank had long modified his bikes to shift gears on the right side, and just stopped bothering about the rear brake, but having bloody toes after each race was getting to be a drag, figuratively as well as literally. Just keeping himself in left socks was a problem. The things couldn't even be darned because of all the clotted gore.

Of course, he could have gone out on the track just to cruise, or put in token appearances now and again as some other old-timers did. The trouble was, Frank didn't know how to cruise. Once he got on the track, he had to *race*. He had to go for it, boogie, keep cooking, ride to win, the way he had for nearly forty years. He'd get all screwed up if he tried to drive slowly. He wouldn't even know where he was on the course. As for token appearances, well, maybe if he had sons to carry on racing he might have done that. Like Yvon DuHamel.

DuHamel had two fast boys, Miguel and Mario. By the latter part of the 1980s both kids started smoking a lot of people on the track, especially Miguel. Old man DuHamel could come to the races, watch his sons, then once in a blue moon blister along the track himself just for fun.

So maybe—though young Dave Hodge, Ken Hodge's boy, was riding vintage as well and it hadn't slowed his old man down or made him skip too many races. The Canadian racing couple Gary and Mary McCaw also kept competing right along with their teenage son, John (who had been a good, fast kid; a great pity that he died of a tragic illness before he could really spread his wings). But maybe if Frank had had a son he would have started taking it easier—who could say?

His daughters did ride motorcycles, and once Puppy actually raised the subject of racing. Frank would have let her race if Puppy had twisted his arm, but he did little to encourage it. He did little, partly because the idea of women racing seemed a bit strange to Frank and partly because he was terrified of Jana. Bad enough that on weekends she was now often left alone in the house because both girls were at the track.

Back in his days hardly any women raced. They were still

rare, but there were women on the track these days and a few were doing okay in the smaller-displacement or less competitive classes. Well, there was really no reason why they shouldn't have been doing okay. Road racing requires some upper-body strength, more than most people imagine, but it isn't beyond the muscular capabilities of reasonably athletic women. The rest is co-ordination, concentration, and guts, which women ought to have as abundantly as men.

Some did; by the late 1970s among every hundred road racers there were about five or six women. Competitors like Terri Weischofer, Yvette Thrush, Gina Bovaird, and Liz Wemett had done well in the United States, among others, while Germany's Margret Lingen finished the gruelling Isle of Man TT, unlike a lot of the men who tried. In Canada, Kathleen Coburn was doing fine, especially in endurance racing, along with her teammate Toni Sharpless, daughter of the old-time BMW racer Bill Sharpless.

Girls on the track often had ex-racer fathers (as did some of the boys), while other female racers were married to competitors. Mary McCaw, a Ph.D. biochemist in civilian life, could pass any number of men in vintage, and when in top form she could even chase her veteran hotshoe husband Gary McCaw. Pretty Nancy Morrow could also dust many guys in her lightweight classes (though she'd probably have found it easier to catch her husband Brian Morrow off the track than on it). Nurse Sharon Bowyer, Paul Bowyer's wife, was an enthusiastic racer, as was Pro Formula Two competitor Jim Struke's wife, Alicia. Frank especially admired a certain petite, young girl who could rarely finish a race because she almost always crashed, sometimes right on the starting line, yet who'd always come back to the grid the following week. "I like her," said Frank, quite seriously. "Yeah, this broad more cuckoo than me."

Besides, why shouldn't women race? After all, Canada's most successful racer, a person who had once been among the top ten in the world, was a woman. She was *now* a woman, anyway, whatever she may have been before. To say that Frank did a double-take when he found out about it would be an understatement. He simply couldn't believe it, he thought people were pulling his leg, but it was evidently true.

Mike Duff hadn't simply retired from racing. He hadn't simply started taking walks in the woods and photographing birds. He was now called Michelle Duff. Mike had become a girl.

Well—Frank was from the big city, he had seen a lot of things in his life, he was no stranger to the ways of the world, but this bit of news rendered him speechless. He could have believed anything rather than this. Why, he had known Mike since Brno. He had seen him on the track, he had played soccer with him in Austria, he had chatted with him a million times. He had met both of his *wives*. He had admired the guy, both as a competitor and as a close personal acquaintance. He was grateful to Mike for helping him when he first came to Canada, and now Mike had gone and done this.

Frank fully realized that whatever prompted anybody to turn into a woman it wasn't to spite or upset him, but he still felt betrayed. Maybe he was old-fashioned, but if Mike Duff could become a woman, there were just no rules anymore. Anything was possible. Things were truly falling apart, the centre didn't hold, and mere anarchy was being loosed upon the world. Yes, the blood-dimmed tide was being loosed and everywhere the ceremony of innocence was being drowned—and though Frank had never read W. B. Yeats, at the moment he learned about Michelle Duff he would have agreed with all of the Irish poet's sentiments.

Well, okay. People did what they had to do. Good luck to Mike, Michelle, whatever. He or she had been fast once and nothing could erase that.

As for Frank, he would go on doing what *he* had to do. He'd go on doing it even more grimly. He would do it even more stubbornly, because he'd be doing it not just for himself but for all the old ways, all the old feelings, all the old guys who had crashed or had given up or had gone funny. If he was going to be the butt of a thousand jokes, the weird old Boogieman, "Crazy Frank, Part XXXXII, etc.," as a sportswriter put it once, so be it. He'd be doing it for the kids he had taught how to race back in the seventies (and who were now retired) and for the guys who had taught him in the fifties and who were now dead. He'd be doing it for Gustav Havel and for Gerhard Mitter and for Jirka Kostyr and for Standa Malina and for Gary Hocking, and yes, also for his one-time friend, the fastest Canadian, the world championship rider, Mike Duff.

In the early 1970s it occurred to a Toronto businessman named Jack Burnett that perhaps he should start riding motorcycles. He had been racing cars and playing polo rather keenly all his life, and now he was looking for something on which to tool around in the street. He ended up buying a handsome Norton Commando, one of the last models of the classic British marque, but when Mrs. Burnett looked at it she felt that there ought to be some limits to risk-taking in the family. Rather than argue the point, Burnett parked the Norton in his garden. It was still sitting there thirteen years later when George Jonas, a somewhat short-sighted man, happened to stumble over it.

"Look, an old Norton," said Jonas perceptively.

Jonas knew what the rusty object was because some thirty-four years earlier he had shared Frank Mrazek's passion, though not his talent. Jonas gave up racing motorcycles by the time he turned eighteen and occupied himself mainly with writing books: a less hazardous occupation in some ways, if not in others. However, he vaguely kept up with the sport, still rode in the street sometimes, still gawked at photographs of current champions in the newspapers, and had a lingering affection for old twins with wire wheels. He thought that it would be fun to own one, but just then there were other things on his mind so he let the notion pass.

The writer had a friend named Terry Wolfe, a mechanic and

bike-builder who had already restored a Norton Commando for himself. Wolfe had turned his own machine into a rather beautiful yellow thing, the replica of a works production racer, and the minute Jonas saw it he was reminded of the sad old Norton gradually becoming part of the eco-system on the Burnett property. Wolfe expressed some interest in rescuing one more relic, and since Jack Burnett was out of the country at the time, it fell to his brother Ted to facilitate the abduction. Ted's view was that Jack would probably not even notice that the Norton was missing.

The prediction turned out to be correct. However, when the restored Norton was unveiled for Jack about a year later, he generously remarked that by tradition the prince who awakened Sleeping Beauty could marry her. Jonas took this to mean that the Norton now belonged to him and he quickly hustled it away.

The next link in the chain of events was Harry Strothard, then president of the Ducati Owners Club of Canada. Strothard saw a glossy photograph of Wolfe's handiwork in a magazine, and he mistook the restored Norton for a quasi-racing bike. It was true that, cosmetically at least, the Norton now looked like one. Strothard also mistook Jonas for a racer, and invited him to run the bike at the Ducati Club's next meet at Grattan Speedway in Michigan.

Strothard's second mistake was harder to understand because Jonas didn't look anything like a racer, even cosmetically. He looked more like a turnip. Common sense suggested that an aging writer who resembled a root vegetable should say thanks but no thanks to Strothard, which Jonas promptly did; but then common sense deserted him. He telephoned Wolfe. A couple of weeks later the writer and his mechanic found themselves leaning their Nortons cautiously into Turn 1 at Grattan.

For Wolfe it was the first time. At thirty he was a seasoned competitor in the dirt, but had never raced on pavement before. For Jonas, fifty-two, it was the first time in three and a half decades. Though Strothard had assured him that this wasn't going to be a race, just a club rally for a bunch of old ladies, the people whizzing by on both sides looked neither old nor particularly ladylike to Jonas.

Just about everybody passed him at Grattan, but one person kept passing him as if he were standing still. By the end of five laps this hotshoe must have lapped the turnip-ridden Norton three times. The myopic Jonas could barely catch a glimpse of his number plate, 77, as he kept buzzing by on the outside.

About a month later, a sucker for punishment, Jonas returned for some further humiliation to the vintage events at Mosport. In the paddock he saw a stocky, greying gentleman being assisted by two pretty teenage girls onto a wicked-looking Triumph triple. One nubile was holding the bike, while the other attempted to swing the rider's leg over the saddle. It was an intriguing sight and Jonas approached for a closer look.

The number plate on the Triumph read 77. As for the greying gentleman—well, for a split second Jonas thought that he resembled former North American Superbike Champion Frank Mrazek. Then he shrugged off the notion, ascribing it to his short-sightedness. It was 1987, and Mrazek must have retired ages ago.

Jonas, of course, turned out to be mistaken. When he started chatting with Mrazek, first at the track and later over dinner, he was struck by a sober remark the former Superbike champion made about his career.

It wasn't that Mrazek was shy or modest about his racing achievements—on the contrary—but after thirty-eight years he could view them in perspective. He knew that they fell short of his promise and his expectations. Maybe they fell short by only a step, except it was the only step that mattered. It was the step that separated those who were inside the winner's circle from those who remained outside. "Some guys I used to smoke," said Mrazek, "today are in this book that I read about. I guess nobody write in book about me now, but maybe somebody still give me bike. If I get bike, I go boogie in Daytona."

This was Mrazek's remark, and it gave Jonas an idea. In fact, it gave him two ideas.

In life everything is a matter of luck. It was pure luck that Frank was now sitting on a 1,000 cc Ducati: a magnificent ex-Daytona Cup-winning Ducati, U.S. Team Leoni ace Jimmy Adamo's spare bike, and probably the sharpest BOTT weapon in Canada. It was pure luck that a bunch of strange guys—lawyers, businessmen, film directors, fashion photographers, and God knows what else—who had had absolutely nothing to do with motorsports before suddenly decided to go racing. Inside less than a year they had set up a team, Classic Racers Canada, then handed Frank the big Ducati and said: okay, go boogie. It was pure luck.

Frank was a bit dubious at first when Jonas, who seemed the most peculiar of the lot, started dragging him to all kinds of restaurants to meet this fellow and that fellow. It all sounded like a load of bull, but what was there to lose? Nineteen-eighty-seven was rock bottom. If nothing came up, he just wouldn't be racing next season. Hope was a good dish for breakfast but not much of a dish for supper, yet one still had to feed on it when there was nothing else in the cupboard.

So Frank went and had coffee with a big-shot named Eddy Cogan who had a very pretty daughter, Cari. It was at the restaurant of Cogan's friend, a funny old bird people kept calling Stubby, and Frank was surprised because they were supposed to be all high-powered people in that place, according to Jonas; but they sure didn't look it. They looked like a bunch of ordinary folk wearing their Sunday best at any neighbourhood joint in Mississauga. Well, with the exception of one guy, Dusty Cohl, who was wearing a cowboy hat, of all things. They said he was a lawyer, and Frank thought, really, he looked like anything but. Which was all right; greying, limping, and overweight, Frank didn't look like Central Casting's idea of a motorcycle racer either.

Then Frank met a couple of nice young lawyers, Ted Burnett and Peter Israel, who told him that they had ridden motorbikes, and another lawyer he recognized from the news, Eddie Greenspan, who said that once he had fallen off a motorbike. Well, that was fun with the lawyer showing off his wrists and Frank showing off his ankle, and everybody had a good time. There was some chatter about other important people, including one or two Frank had heard about, like this big film director Norman Jewison, and also a big company, Labatt's, but he still didn't believe any of it until one day the Ducati was unloaded from the van. And it wasn't just a bike, but also a tuner to go with it, maybe the hottest guy for racing Ducatis in the country, Tim Spiegelberg.

It was a Spiegelberg-built Ducati that Frank had been chasing for the last couple of seasons on his Honda in the Eastern Canada Challenge: a very fast 750 cc machine that tuner Tim had put together for his younger brother, Terry. Rumour had it that BOTT champion Larry Strung might not be racing in 1988, so in the Battle of the Twins Frank's biggest competition could be young Spiegelberg, except now Frank would have the bike to keep up with him. (Tim couldn't lose: he'd be tuning both Ducatis.) Then, if Frank could win the Canadian BOTT championship, perhaps the team would try for the Pro Twin GP in Daytona in the spring of 1989.

Daytona!

It was a new lease on things, and it all had to do with people Frank hadn't even known until a few months ago. It showed how everything in life depends on chance, though what is good luck for one person could easily be bad luck for another. For instance, Jana was an honest person; when she first met Jonas and realized what the writer was trying to organize for Frank, she just looked him straight in the eye and said, "You're my enemy."

It was also a matter of chance, except ill chance this time, that the 1988 season started off with three crashes even before the new team had acquired Frank's Ducati. In March, while setting up his own bikes for the Daytona vintage races, Frank highsided the Honda at the Roebling Road track in Savannah, Georgia. The spill separated a muscle in his shoulder and there was nothing to do about it except live with the pain: that's what ice and 222s were for. Then he crashed twice at Daytona, but there his luck was better. When his front brake failed and he had to lay down the Honda in the greenery he was just coming out of the infield dogleg: a left-hand sweeper where even vintage speeds can hit about ninety to a hundred mph—yet Frank just slid for a while, then picked himself up and walked away from it. Frank's daughter, Puppy, saw the crash and stopped breathing until her father got back on his feet, then she grinned and ran to fill another bucket with ice.

The second Daytona crash was a low-speed slideout at the infield hairpin, the International Horseshoe, as Frank was dicing for third place in the 500 Premier class. He was trying for an inside pass at a spot which he had previously warned Jonas to stay away from because it was too slippery—but since there seemed to be an opening, Frank went for it. By the time he restarted the bike the entire field had disappeared, so he had to chase to catch them up and finish eighth. Then he took the Sleeping Beauty Norton and piloted it to fourth place in the 750 GP class. (Later on the same bike Jonas was happy to manage forty-second in the much slower 750 SP class.)

And with that the 1988 season was on its way.

Nearly seven months later, on the 25 of September, the season was over. Frank, now fifty-two, was a champion once more. He had won the Canadian title in the Battle of the Twins, and for good measure he also nailed down the VRRA 500 cc Vintage Championship. Altogether he won fifteen races out of thirty-two starts.

It wasn't as easy as it sounded. The main BOTT competitor turned out to be the veteran Paul Bowyer, who put up a stubborn resistance aboard his old Norton. Though Frank wasn't the underdog for once, the championship could have gone either way until the final race.

Classic Racers Canada had its hairy moments, like most teams, including three crashes. Once Terry Wolfe stuffed it into the rhubarb at Shannonville's Corner 2, but only lost some skin which he could re-grow at no expense to the team.

Frank, however, banged himself up pretty badly at Loudon, N.H. He broke four ribs when the spokes on the rear wheel of a competitor's bike collapsed in front of him, resulting in a fiery, multiple wreck that cancelled a round in the Am-Can Supervintage Challenge. This was in July. Then in August Frank finished what he had begun at Roebling Road by separating the rest of his shoulder when he lost the front end of the Ducati just before Corner 6 at Shannonville.

This accident upset Tim Spiegelberg, who felt that if a racer loses the *rear* end of a machine it's his own responsibility—he has a lead throttle hand—but if the front end slides out it reflects badly on his tuner. Frank, of course, insisted that he couldn't understand what happened because he was only cruising. He maintained this position until confronted with the official timer's evidence: on the lap before his crash he had broken the lap record for BOTT at Shannonville's Nelson short track. Frank replied, without missing a beat, that this proved nothing as the previous lap record had also been his.

The Supervintage Heavyweight title went to Ken Hodge. Frank was fifth of thirty riders in the hunt for championship points and he might have done better if a young man named Larry Rose had appeared earlier on the scene. Larry was a Triumph whiz, and he finally sorted out the ex-Paul Smart triple. Until Larry's arrival the fine, blue-blooded machine had served mainly as an exercycle for the team. People kept loading and unloading it from the van, then pushing it up and down pit lane, and Jonas calculated that by mid-season they could have pushed it all the way back to Britain.

What Jonas found a bit harder to figure out at the end of the season was why people were doing what they were doing. It certainly seemed to fly in the face of everything from Adam Smith to Karl Marx.

For instance, why would young craftsmen like Rose or Wolfe work all summer for nothing just to make some ancient bikes function and maybe finish some unheralded races? Was that a smart way to maximize profits? Did it score points in heaven?

What about the sponsors? Why would sober, successful citizens most of whom had never been to a motorcycling event put up hard cash so that an old guy they had never heard of might win one more title in a sport about which they cared nothing? Speed contests had no redeeming social value, supposedly. They were not some heavy-duty, fashionable, recognized charity. What was the name of the altar for this burnt offering? Was it a small sacrifice to defiance? Or, perhaps, to courage?

People often said: Well, games are games. They're sports. They all involve plenty of hardship, self-denial, and skill, so speed contests are no different. This was quite true, provided one could see no difference between gladiators and athletes, and by extension no difference between life and death. (Which, of course, is possible on some higher cosmic plane.)

Or, equally puzzling, why did speed contestants put their lives on the line to finish a position ahead of somebody else, then negate it all by helping an opponent to finish ahead of *them*? Because that's what racers were doing all the time. At Loudon, for instance, Gary McCaw began by lustily chasing Mrazek, then spent half the night patching up Mrazek's bike after his crash so he could chase him again in the morning. McCaw might have had Mrazek's position the next day just by doing nothing. How did that fit in with enlightened self-interest? Another rider named Brian Kenyon threw away his own race to pull a bike off a competitor who had crashed. And in the BOTT championship it was Paul Bowyer's wife Sharon, an operating-room nurse, who taped up Frank's shoulder so he could make the grid. Sharon's husband may well have won the title if only Sharon had not lifted a finger that day, but people who did not lift a finger to help their rivals were the exception at the track, not the rule.

Why?

What did it say about so-called blood sports? Gentle members of the caring professions, for instance, rarely seemed as co-operative with one another. In a competition involving life and death Jonas would have felt less secure among a bunch of social reformers or lyric poets.

Were competition and co-operation the polar opposites people often set them up to be? Or were they rather two sides of the same coin?

What were the myths of the start/finish line? What did people really believe in? Tim Spiegelberg, for instance, was a cool, young scientific type, fond of slide rules and flow-benches, but during a race he rarely uncrossed his fingers. The starter's flag would drop, and Spiegelberg's fingers would snap at right angles and stay firmly crossed until the chequered flag. He wasn't aware of it and looked somewhat sheepish when it was pointed out to him, but he kept doing it all the same. Just as Frank would never appear in the paddock without his lucky straw hat.

On the whole, people on the track were not mystics any more than they were humanitarians. They were competitors, high priests of speed, hardheaded devotees of the industrial arts. What deity did they invoke with lucky headgear and crossed fingers?

Jonas was puzzled. The closest he could come to an explanation involved such general phrases as the human condition or the human spirit. They were just words that answered nothing, but they may have been true for all that.

The hillside is dotted with small fires. The mosquitoes—well, somebody's forgotten to show the mosquitoes the calendar. In these parts they're not yet supposed to be buzzing around in March, so they're a month or two ahead of schedule.

The small fires on the hillside are privateers, barbecuing steaks or wieners. The veterans, the real connoisseurs, are probably barbecuing sausages. There's zilch catering at this track but the sausages in the village are great. They're almost as good as the hot sausages at Daytona, and the veterans know it.

There are no factory guys around the fires. The factory guys with their wives or girlfriends are sleeping in a motel. At Daytona they'll be sleeping in Daytona Beach or maybe in Ormond Beach.

Daytona. Well, there's always an outside chance. Other privateers have won it, including people no faster than you, so why not you for a change? Anyway, no one expects you to win it. It's enough to be there. Maybe finish in the top five.

Other people have finished in the top five. True, most were a bit younger. So what? They couldn't help being younger, just as you can't help being older. You were younger once, and all these juniors will be older one day, too.

Old age is like any other misfortune, it brings out the good and the bad. Getting older is just like being in trouble: it makes good guys better and bad guys worse. Maybe, with luck, you're one of the good guys.

You'd better be, because nobody remembers. In racing there is only today. Nothing lasts forever, but champions don't even last a blink of an eye. Today the same kids who would genuflect or maybe faint if Eddie Lawson sauntered by, might not even twig at the name of Mike Hailwood or Jarno Saarinen. To say nothing of Freddie Frith.

Which is okay; it only means that you've got to keep cooking.

People say you're crazy. They say it so often that you begin to wonder: yet what is sane about being scared? Is that what sanity means, being scared?

Because all these guys who call you crazy are scared. Yeah, scared—not just of crashing or pain, but scared of the boss, the bank, or their neighbours ending up with a bigger house. They're even scared of their girlfriends, some of them. You name it, they're scared. Then they tell you, "You're crazy, I always do what I want"; but how can any guy be free when he's scared?

It's not a question of racing. Hell, a lot of guys wouldn't get a kick out of racing. But when a guy is not doing whatever it is he wants to do because he's worried that it's crazy: *then* you know he's scared. That's what being scared means.

And some of them also say, "Well, I want to live long, die a natural death". Okay, and isn't that crazy? How do you live long without becoming piteously old? Maybe so old you can't scratch your ass or tie your own shoelaces? What's so sane about that?

Have these sane guys ever been inside a place for folks who've lived long? Have they seen what natural death looks like? At least you know what violent death looks like. That's what it looks like: you're puzzled for a second, then go out like a light. It feels the same whether you wake up after it or not.

True, you can get scrunched up and go on living like a vegetable, but that can happen anyway. It can happen when you fall down in the bathtub or if you live long and peacefully enough for a whacking big stroke. Either way, you take your chances.

As long as you're not scared, you're free, and as long as you are, well . . .

Zap another mosquito.

Maybe that's what is crazy: to want to be free. A lot of people wouldn't cross the street for it. Anyway, those fires are still burning on the hillside. Tomorrow at the track you set up the bike; the day after, Daytona. Nobody, but nobody ever twisted your arm.

POSTSCRIPT

On March 10, 1989, Frank Mrazek, fifty-three, riding Classic Racers Canada's Ducati, made the 4th row on the starting grid of Daytona's Pro Twin GP fifty-mile event. The Ducati's rear wheel seized on the first lap and he was out. That's racing, but it's not the last lap yet. At fifty-three, Mrazek is planning to defend his Canadian title in the Battle of the Twins again.

INDEX